Holocaust Education – Historisches Lernen – Menschenrechtsbildung

Series Editors

Anja Ballis, Institut für Deutsche Philologie, Ludwig-Maximilians-Universität München, München, Germany

Michele Barricelli, Historisches Seminar, Ludwig-Maximilians-Universität München, München, Germany

Markus Gloe, Geschwister-Scholl-Institut für Politikwissenschaft, Ludwig-Maximilians-Universität München, München, Germany

Die Reihe „Holocaust Education – Historisches Lernen – Menschenrechtsbildung" verbindet inter- und transdisziplinär die beiden Ansätze von Holocaust Education und Menschenrechtsbildung, die sowohl im Bereich der Gesellschaftswissenschaften, der Sprachwissenschaften als auch im erziehungswissenschaftlichen Gesamtkontext der Vermittlung von demokratischen Werten in bildungspolitischen Zusammenhängen adressieren. Ausgewiesene Expertinnen und Experten aus verschiedenen Disziplinen, aber auch der wissenschaftliche Nachwuchs präsentieren in dieser Reihe neueste Forschungsergebnisse, theoretische Grundlagen und dokumentieren die aktuelle inter- und transdisziplinäre Diskussion. Der wissenschaftliche Beirat der Reihe setzt sich zusammen aus Prof. Dr. Sascha Feuchert (Justus-Liebig-Universität Gießen), Prof. Dr. Jeanette Hoffmann (Freie Universität Bozen), Prof. Dr. Martin Lücke (Freie Universität Berlin), Prof. Dr. Tonio Oeftering (Carl von Ossietzky Universität Oldenburg), Prof. Dr. Martin Rothgangel (Universität Wien) und Dr. Noah Schenker (Monash University, Melbourne). Die Reihe „Holocaust Education – Historisches Lernen – Menschenrechtsbildung" wendet sich an Wissenschaftlerinnen und Wissenschaftler, die sich mit Fragen der Vermittlung des Holocausts und Fragen der Menschenrechtsbildung beschäftigen, sowie historisch-politische Bildnerinnen und Bildner in Schule und außerschulischen Kontexten.

More information about this series at https://link.springer.com/bookseries/16330

Anja Ballis
Editor

Tour Guides at Memorial Sites and Holocaust Museums

Empirical Studies in Europe, Israel, North America, and South Africa

 Springer VS

Editor
Anja Ballis
Institut für Deutsche Philologie
Ludwig-Maximilians-Universität
München
München, Germany

ISSN 2662-1878   ISSN 2662-1886 (electronic)
Holocaust Education – Historisches Lernen – Menschenrechtsbildung
ISBN 978-3-658-35817-4   ISBN 978-3-658-35818-1 (eBook)
https://doi.org/10.1007/978-3-658-35818-1

Responsible Editor: Stefanie Laux
This Springer VS imprint is published by the registered company Springer Fachmedien Wiesbaden
GmbH part of Springer Nature.
The registered company address is: Abraham-Lincoln-Str. 46, 65189 Wiesbaden, Germany

Contents

Editor and Contributors

About the Editor

Anja Ballis Professor, Chair of German Language Education at the University of Munich. Research interests: Holocaust education, educational media research, qualitative research.
https://www.germanistik.uni-muenchen.de/personal/didaktik/professoren/ballis_anja/index.html

Contributors

Anja Ballis Professor, Chair of German Language Education at the University of Munich. Research interests: Holocaust education, educational media research, qualitative research.
https://www.germanistik.uni-muenchen.de/personal/didaktik/professoren/ballis_anja/index.html

Franz Breuer 1980–2014 Professor of Psychology at Westfälische Wilhelms-University Münster. Research interests: reflexive grounded theory methodology, stages in aging, family transfers between generations.
www.grounded-theory.net/

Sonja Danner Doctorate in Religious Education, Educator at the Private University College of Teacher Education Vienna/Krems and at the University of Vienna. Research interests: learning concepts of memorial education and Holocaust education.
https://ufind.univie.ac.at/de/person.html?id=63599

Brenda Gouws Doctorate in Education, honorary research affiliate at the Kaplan Centre for Jewish Studies Cape Town, academic editor. Research interests: narrative inquiry, history teachers, personal stories, history education, Holocaust education.
www.researchgate.net/profile/Brenda_Gouws

Franziska E. Müller student at Europa University Viadrina, Faculty of Law. Former volunteer at Dachau concentration camp memorial site. Research interests: international law, Holocaust (education).

Moritz Lautenbach-von Ostrowski Doctorate in German Linguistics at the University of Hamburg, since 2019 teacher in social work. Research interests: forms of communication in institutions, and the relationship between language and memory, especially with regard to National Socialism; oral history, theories of memory.
https://uni-hamburg.academia.edu/MoritzLautenbachvonOstrowski/Curriculu mVitae

Michael Penzold Doctorate in German Literature, senior researcher in German Language Education at the University of Munich. Research interests: teachers and the Holocaust/Shoah, literature and media in Holocaust education.
https://www.germanistik.uni-muenchen.de/personal/didaktik/mitarbeiter/pen zold1/index.html

Irene Ann Resenly Doctoral candidate at University of Wisconsin-Madison, Curriculum and Instruction & Social Studies Education. Research interests: Holocaust education, teacher education, German collective memory.
resenly@wisc.edu

Inese Runce Doctorate in Social Science, leading researcher at the Institute of Philosophy and Sociology at the University of Latvia, lecturer in Baltic history, cultural and religious studies at the Faculty of Humanities at University of Latvia. Research interests: relations between state and church, history of Church in Latvia, regional and religious studies, Jewish studies.
www.fsi.lu.lv/?sadala=152

Emilia Smechowski Journalist, columnist and author of the book *Wir Strebermigranten* (2017) and *Rückkehr nach Polen* (2019). She was awarded the German-Polish Tadeusz Mazowiecki Journalist Prize, the German Reporter Prize, and the Konrad Duden Journalist Prize.
https://twitter.com/emilia_owski?lang=de

Aija van der Steina Doctorate in Social Science, senior researcher of the Institute of Philosophy and Sociology at the University of Latvia, visiting lecturer at the Tourism Masters' Program at Vidzeme University of Applied Sciences (Latvia) and the Graduate Tourism Program of Monash University (Australia). Research interests: difficult heritage and tourism, destination management and marketing, tourism planning and impact, diaspora tourism.
https://independent.academia.edu/AvanderSteina

Johan Wassermann Professor of History Education and Head of the Department of Humanities Education of the University of Pretoria. Research interests: youth and history, history textbooks, teaching controversial issues, institutional biographies of higher educational institutions, minorities and the minoritized in colonial Natal.
www.up.ac.za/humanities-education/article/50813/staff

Julie Winter Doctorate in German Literature, Visiting Professor of German at Western Washington University, literary translator. Research interests: translation studies, stylistics, intercultural memory studies.
https://chss.wwu.edu/winterj8

Tour Guiding at Memorial Sites and Holocaust Museums

Anja Ballis

1 Commonalities in Diversity—The Tour Guide's Role

Tour guiding at memorial sites and Holocaust museums offers a wide range of possibilities for connecting visitors with history, commemoration, and education. In this volume, contributors reflect on how to teach and mediate difficult history from the perspectives of guides. The authors of the articles cover different countries and continents, different educational systems, and different levels of involvement in the history of the Holocaust.

Investigating guides at Holocaust museums and memorial sites makes us aware of the diversity of the guides and the varieties of their tasks. Different biographical backgrounds and motivations characterize this group and their practices of arranging tours (Ballis 2018). Although the differences are considerable, I want to highlight the commonalities of tour guiding at different places of memory.

On the one hand, tour guiding depends to a great extent on individuals' willingness to commit themselves to the goals of the institution and to pursue their activities on a part-time and voluntary basis (Zumpe 2012, p. 90). At museums and memorial sites, guides dedicate their time and activities to ensure that the discussion about Holocaust and Nazi crimes is not limited to experts in institutions. In light of history and its mediation, they keep memory practices alive; further, they contribute to the public debate about whether and how genocides can be prevented in our time. Each guide offers a unique service at the concentration camp memorial, by sharing a personal interpretation of the subject matter

A. Ballis (✉)
Institut für Germanistik, Ludwig-Maximilians-Universität München, Munich, Germany
e-mail: anja.ballis@germanistik.uni-muenchen.de

© The Author(s), under exclusive license to Springer Fachmedien Wiesbaden
Gmbh, part of Springer Nature 2022
A. Ballis (ed.), *Tour Guides at Memorial Sites and Holocaust Museums*,
Holocaust Education – Historisches Lernen – Menschenrechtsbildung,
https://doi.org/10.1007/978-3-658-35818-1_1

with visitors. In doing so, they teach members of civil societies about the history and about the site; vice versa they carry the history back into the society they belong to. As can be seen in various articles of this volume, guides feel either well qualified to perform these tasks or they feel left alone and overburdened by them.

On the other hand, guides at museums and memorial sites face challenges associated with meeting visitors. The activities of tour guides require certain demands from them; besides learning historical facts about the Holocaust, they have to figure out how they fit in at their institutions. Although they have different professional backgrounds, all guides face the same communicative challenge: They have to convey the cruelty and inhumanity of the events to visitors having different intellectual levels and emotional states (Österberg 2017, p. 266). To fulfil this task, guides need factual knowledge and empathy with their groups. Communicating with people is a fine art that requires experience.

In addition, the role of tour guides is in flux as a consequence of digitization. The rise of technology has also had an impact on exhibits in museums and on navigation at sites (Hoskins 2017). Currently, we find all over the world the tendency to modernize Holocaust exhibits: Apps and Virtual and Augmented Reality tools are being integrated. Many of these tools enhance the visitors' abilities to tour the site on their own, and they support the experience of different historical layers on site. Furthermore, many tools prompt users to interact with other visitors by sharing what they have seen (Feldman and Musih forthcoming). These new media formats challenge tour guiding, since they partly compete with the service of the guides and partly replace them. After the Covid-19 pandemic, it will become clear what role guides will take in light of the abundance of online offerings.

Some of the institutions mentioned in this volume were established by former victims. At some places, the former inmates pushed initiatives to build a memorial site or a museum where people can learn about the history of the Holocaust (Seiter 2017). Survivors have also supported the educational work at the institutions by guiding people through exhibits and grounds. It is important to consider their role as facilitators, too. In this context, we have to understand the victims both as subjects of shaping history and "objects" shaped by institutions. The close bond with survivors can still be felt today at memorial sites and museums. Both visitors and guides might descend from survivors, which could evoke a special atmosphere on site.

Since the sites are regarded as places of education, experts at museums and memorial sites have developed programs for teaching and learning. The institutions mentioned in the volume provide a variety of pedagogical programs for their visitors. In museums, the teaching of the Holocaust is often located in the context of human rights education. Reflecting on universal human rights, guides might outline connections between the past and the present. At other institutions, guides teach historical aspects mainly to high school students. Guides provide historical knowledge at these authentic places where history "really" happened and where visitors might feel the "breath of history" (Gussmann et al. 2019, p. 183–185). Finally, there are some institutions without any pedagogical programs. Particularly at smaller sites, often sub-camps of former concentration camps, visitor groups have to bring their own guide to learn the history of the place.

Guides find different starting points for their educational work. They either modify the programs offered on site, or they develop their own tour. Most of the programs at the institutions covered in the volume are geared towards high school students. Students are an important target group, since they are considered the next generation that is expected to be responsible for preventing genocides.

Guides represent an important, often forgotten group of educators. The contributors to this volume show from different research perspectives that it is worth understanding more about the guides' personal interests, their motivations, and their concept of guiding. To this end, authors apply methodologies from the social sciences to describe the guides' point of view based on interviews and ethnographic observations. One researcher uses social science methods to understand a novel, thus highlighting the often forgotten group of guides in a poetic setting. Another unique category is the work of an author who portrays guides' activities with a literary montage, using social sciences methodology. Complementing the various approaches in tour guide research, a detailed linguistic analysis sheds light on a survivor's testimony echoed in the guides' language.

It is worth mentioning that in the various articles of this volume, the authors use different terms for "guides." This term serves here as an umbrella term for the activity of accompanying visitors through exhibits in museums and the grounds of memorial sites, informing them about historical events along the way. Depending on the specific character of the activities at the particular institutions, the terms docent (especially in the US and Canada), museum's educator, and teacher guide are commonly used.

2 The Structure of the Volume—Global Perspectives

All over the world, we find people who dedicate their time and skills to keep history alive. The volume reflects a global perspective, examining museums and memorial sites in Europe, North America, and South Africa in greater detail.

However, we must critically note that contributions from other parts of the world, like South America, India, and the Far East, are missing. Nevertheless, the studies gathered in this volume open up an orientation for further approaches of tour guiding based on and centered around "authentic" materials from guides.

Beginning with Germany and Austria, the first section of the volume brings articles together that focus on the role of guides and trace their efforts of professionalization in the institutions. In the article "Guides at Memorial Sites and Holocaust Museums—Professionalization in Times of Change in the Culture of Remembrance," I investigate two tendencies of tour guiding nowadays: First, exhibits are being redesigned, in particular to incorporate novel, media-based presentation formats. Second, there is a discernible trend toward professionalizing—and often also standardizing—the training of educators and guides.

Based on field studies, interviews, and document analyses at four institutions—the memorial sites at Dachau and Mauthausen, the United States Holocaust Memorial Museum in Washington, and the Anne Frank Zentrum in Berlin—I try to answer the question about which core activities are expected of guides during tours at memorial sites and Holocaust museums.

In her article on "Site Educators in Germany's Perspectives of Practice. The Sense-Maker and the Storyteller," Irene Ann Resenly points to a similar direction. She focuses on memorial sites as sites of learning. In a case study at a German institution, she interviewed two site educators to highlight the role that they play for visitors. Of special interest are their dynamic and purposeful pedagogical decisions; the author explores how memorial site educators conceptualize Holocaust education and to what extent notions of history, memory, and place shape their ideas.

With the help of linguistic analysis, Moritz Lautenbach-von Ostrowski reflects on "Passing Down Testimony—How Concentration Camp Memorial Guides Meditate Testimony Through Linguistic Action." Based on a survivor's story at the Bergen-Belsen camp, the manner in which a guide conveys this testimony is examined in detail. In three case studies, the author comparatively shows how the "textualization" of oral testimonies is reflected in language and consolidated by third parties. Further, the study explains how linguistic analyses are useful tools for interdisciplinary Holocaust research and provide a contribution to cultural studies.

Since educating high school students is an important goal at memorial sites and Holocaust museums, a special type of a teacher guide is the focus of two articles.

Michael Penzold considers in "Informing, Accompanying, Commemorating at the Memorial Site Dachau—Teacher Guides' Experiences and Reflections" the process of teachers becoming tour guides at the Memorial Site Dachau. The dynamic process of self-reflection among the teacher guides is outlined in three stages: changing roles, developing language and behavior routines, and personal engagement. Based on qualitative interviews and ethnographic observations, teacher guides continue to ask questions and struggle for individual perspectives throughout all three stages.

In addition, Sonja Danner takes a look at the approaches used by teachers in different subjects to Holocaust education. She interviewed teachers who had accompanied their high school students to places of Holocaust remembrance. In her article "Teachers as Guides at Memorials and Places of Remembrance," she argues for a participatory approach to visiting those sites and stresses the role of subjects and personal involvement. She recommends more intensive engagement, both for students and teachers, when learning about the Holocaust.

In the second section of the book, camps in Eastern Europe are of special concern. The writer Emilia Smechowski reflects on guides in her article "The Auschwitz Concentration Camp—A Journey into Terror." This piece was originally printed in a German magazine and was translated by Julie Winter, who specializes in literary translation of memoirs. Emilia Smechowski conducted in-depth interviews with 12 of the tour guides at the Auschwitz-Birkenau Museum and Memorial. She wanted to understand the nature of the guides' work, how they handle the masses of visitors, and how they are trained and prepared for this demanding job.

Of particular interest is the article "Guiding at the Jewish Holocaust Sites in Riga—Difficult History, Tourism, and Individual Experiences." Inese Runce and Aija van der Steina discuss the formation and transformation of Riga Jewish heritage and Holocaust memorial sites since the fall of the Iron Curtain. They consider the special circumstances of Latvia after the Second World War and how these conditions effect guides' professional work today. Their conclusions are based on theoretical studies and survey results from guides.

The third part of the volume addresses Israel. Franz Breuer's contribution "Identificatory Trajectories of Holocaust Memorial Site Guides—A Theory Sketch Grounded in Yishai Sarid's Novella *The Memory Monster*" analyzes the novel by means of social science methodology. Yishai Sarid depicts the experiences of a young Israeli historian who guides tour groups—mainly Israeli high school students—through the former Nazi German concentration and extermination camps in Poland. Adopting a grounded theory research approach, the article uses the text as the data basis to sketch the guides' identities and how they change over time.

The fourth and final part of the volume covers museums in the US, Canada, and South Africa. The investigated programs at the museums are characterized by a close connection between teaching about the Holocaust and topics of human rights education, especially when addressing younger visitors. Beginning in Canada, Franziska Müller discusses in her article "Self-Guding, Moderating, Accompanying, and Survivor-Guiding—Educational Approaches at the Sarah and Chaim Neuberger Holocaust Education Centre in Toronto" different approaches to guiding at the Neuberger: self-guiding, moderating, and accompanying. Since the Neuberger is not a large museum with numerous guides, its pedagogical program includes a discussion of the Historical Thinking Concept (HTC) and digital educational resources. Lastly, her paper takes into account yet another way of guiding—survivor-guiding. In this context, the focus is on Pinchas Gutter, a survivor who lives in Toronto and often speaks at the Neuberger.

Digitization is an important mechanism for developing new approaches to self-guiding. In my article "The Impact of Digitization on Tour Guiding—A Case Study on Interactive Biographies in Museums" I consider tour guides' perspectives of media innovations. Of special concern is to what extent the docents' understanding and practices of tour guiding at Holocaust museums vary because of technological innovations integrated into the exhibits. I investigated guides' practices with the project Dimensions in Testimony, which the USC Shoah Foundation has been developing since 2011. The study evaluates how the institution trains the docents to use the interactive biographies, and how the docents assess this digital tool in terms of their tour guiding. I used a qualitative approach from the social sciences, conducting interviews and observing the guides' practices.

In the last article of this volume, Brenda Gouws and Johann Wasserman give insights on "Guiding at the Durban Holocaust and Genocide Centre, South Africa." The approach at the center is dedicated to teaching about the Holocaust in South Africa and also about other genocides, and stresses education as its priority. To better understand guiding at the center in Durban, the researchers examine three narratives: the narrative of the place, the narrative of Holocaust education, and the narrative of those who do the guiding. The guides are both professional representatives of the institution and individuals with their own stories to tell. The rationale behind teaching the Holocaust in post-conflict environments is that it enables a more open, objective, less emotion-filled examination of a country's difficult past. This is certainly true for post-apartheid, post-colonial South Africa, where violence, intimidation, fear, discrimination, xenophobia, and murder are seared into the country's collective consciousness.

The authors of this volume provide a number of topics ranging from the past to the present; with the help of various methodologies, the articles offer insights into guides' activities at Holocaust museums and memorial sites in many parts of the world. It is our hope that the authors have been able to shed light on and underline the important and demanding work the guides do every day. If so, then they will have made a contribution to the guides' efforts to create a better world.

Many colleagues supported this publication. My special thanks include Noemi Meixner, Mira Schienagel, and Marianne Wischer, who did the proofreading. Julie Winter provided support in English language, which was characterized by high accuracy and professional seriousness.

I am obliged to Springer Publishing House for printing and distributing the volume.

Munich, Fall 2021

References

Ballis, A. (2018). Confronting subject matter education with memorial pedagogy. Guides at memorial sites and Holocaust museums. *RISTAL*, 1, 19–34.

Feldman, J., & Musih, N. (forthcoming). Israeli Memory of the Shoah in a Digital Age: Is it Still "Collective"? In *Die Zukunft der Erinnerung* (177–192). Berlin, Boston: De Gruyter.

Gussmann, M., Merkt, M., & Schwan, S. (2019). Zur Wahrnehmung und Wirkung historischer Orte. Eine kognitionspsychologische Perspektive. In A. Drecoll, T. Schaarschmidt, & I. Zündorf (eds.), *Authentizität als Kapital historischer Orte? Die Sehnsucht nach dem unmittelbaren Erleben von Geschichte* (175–187). Göttingen: Wallstein.

Hoskins, A. (ed.) (2017). *Digital Memory Studies. Media Pasts in Transition*. New York, London: Routledge.

Österberg, O. (2017). Visits and Study Trips to Holocaust-Related Memoral Sites and Museums. In M. Eckmann, D. Stevick, & J. Ambrosewciz-Jacobs (eds.), *Research in teaching and learning about the Holocaust. A dialogue beyond borders* (247–272). Berlin: Metropol.

Seiter, I. (2017). *Holocausterinnerung im Museum. Zur Vermittlung zivilreligiöser Werte in nationalen Erinnerungskulturen im Vergleich*. Baden-Baden: Nomos.

Zumpe, H.E. (2012). *Menschenrechtsbildung in der Gedenkstätte. Eine empirische Studie zur Bildungsarbeit in NS-Gedenkstätten*. Schwalbach/Ts.: Wochenschau.

Germany & Austria

Guides at Memorial Sites and Holocaust Museums

Professionalization in Times of Change in the Culture of Remembrance

Anja Ballis

Abstract

Two trends can currently be observed at memorial sites and Holocaust museums: First, exhibits are being redesigned, in particular to incorporate novel, media-based presentation formats. Second, there is a discernible trend toward professionalizing and often also standardizing—the training of educators and guides. Taking these changes as its starting point, the present contribution focuses on the training and activities of guides. Based on field studies, interviews, and document analyses regarding four institutions—the memorial sites at Dachau and Mauthausen, the United States Holocaust Memorial Museum in Washington, and the Anne Frank Zentrum in Berlin—the following questions are investigated: What core activities are expected of guides during tours at memorial sites and Holocaust museums? What position do guides occupy within their respective institutions? What suggestions for the optimization of their training and activities can be derived from these findings? The declared aim of this contribution is to reflect the work conditions of full- and part-time guides in times of profound change in the culture of remembrance.

A. Ballis (✉)
Institut für Germanistik, Ludwig-Maximilians-Universität München, Munich, Germany
e-mail: anja.ballis@germanistik.uni-muenchen.de

© The Author(s), under exclusive license to Springer Fachmedien Wiesbaden   11
GmbH, part of Springer Nature 2022
A. Ballis (ed.), *Tour Guides at Memorial Sites and Holocaust Museums*,
Holocaust Education – Historisches Lernen – Menschenrechtsbildung,
https://doi.org/10.1007/978-3-658-35818-1_2

1   Guides at Memorial Sites and Holocaust Museums—Professionalization and Standardization

For some years now, memorial sites and Holocaust museums have been the focus of attention of a diverse range of visitor groups. Students and youth groups go on field trips to these institutions, as do individual and group tourists. Other visitors include Holocaust survivors and their relatives, representatives of various victim groups, policymakers, and members of the general public. Especially in the summer months, these institutions sometimes turn into sites of mass tourism. The steadily growing stream of visitors from all over the world also has an impact on the respective institutions, which must not only react to it organizationally. Rather, they also become aware of their responsibility to present themselves to an international audience. Besides possibilities for telling the story of the location, or the story of the Holocaust, with display panels, audio guides, and mobile apps, guided tours are still considered to be an effective and the most frequently used form of knowledge transmission. Many institutions also offer other possibilities, such as seminars and workshops, encounters with contemporary witnesses, and work in the archives (Österberg 2017, p. 253).

In this chapter, "guides" is understood as an umbrella term for the group of persons who offer guided tours and educational activities in various languages for visitors of different ages at memorial sites and Holocaust museums. Guides are a heterogeneous group with a diverse range qualification. Before they can start work at the respective institutions, they must undergo training. This professionalization measure is shaped by the desire of the respective institutions to provide visitors with concise information and to ensure that certain standards in terms of content, communication, and objectives are satisfied. Guides are thus faced with a complex and demanding profile of requirements that they must meet when performing their work. However, most guides do not have permanent employment contracts with the respective institutions.

The educational backgrounds and job profiles of guides have rarely been the subject of empirical research. In her analysis of a possible link between human rights education and pedagogical ideas at memorial sites in Germany, Zumpe (2012, p. 91 f.) also investigated the professional training of the guides and educational staff at these sites. According to her findings, these persons are often lateral entrants with different levels of interest in and knowledge of historical content and different pedagogical skills. Some of them work at these locations as educational employees and are responsible for a wide range of topics (exhibis, guided tours, etc.); some are freelancers and mainly conduct guided tours; some are teachers who work in special programs on site; and some work for a limited period on thematic projects. Although the guides are trained by the respective institutions, their preparation for the complex conceptual and pedagogical activity often appears to be piecemeal, at most (Thimm et al. 2010, p. 25). Moreover, certified professional training still seems to be a distant prospect (Werker 2016, p. 173). This is a cause for concern in so far as students and teachers stress the role of the guides for a successful memorial site visit: Guides are expected not only to possess specialist historical knowledge, but above all to have the communication skills to "unlock" the site and to encourage the young people to express their views (European Agency for Fundamental Human Rights 2011, p. 49, 69). A number of studies have addressed the communicative practice of guides: In observations of guided tours at Ravensbrück, Dachau, Neuengamme, and the House of the Wannsee Conference in Berlin, Gudehus (2006) identified numerous substantive and formal commonalities between the narratives at the various German memorial sites; these commonalities indicate a canonization of the narrative, irrespective of location and exhibition. Worth mentioning are the efforts undertaken to standardize the tours at the Mauthausen Memorial Site, where the education department developed a communication-oriented education program. When the guides reflected on these efforts, they stressed two points. First, they emphasized the relationship between the students and their teachers, which had implications for communication on site. Second, they expressed the desire for more support for their work from the management of the institution. Different researchers addressed the contrasts between institutionally shaped communication at schools and at concentration camp memorial sites (Halbmayr and Miklas 2014, p. 107; Haug 2015, p. 282; Meseth 2008). From a linguistic perspective, Lautenbach-von Ostrowski (2015) highlighted the way in which guides reported testimonies of contemporary witnesses during tours with students, and the role that they acquired through this "translation work" (Lautenbach-von Ostrowski 2015, p. 271; also Lautenbach-von Ostrowski in this volume).

2   Training, Job Requirements, Institutions, Sampling, and Research Questions

In this pargraph, I reflect on core activities expected of guides. To address this question, the memorial site at Dachau was taken as the starting point of data collections. The selection was based on theoretical sampling, a strategy inherent in grounded theory. In beginning with the Memorial Site Da, an institution was chosen that was trying to professionalize its education program and, in doing so, favored a cognitively oriented approach. At Dachau, guides are referred to as

Referentinnen (f) or *Referenten* (m) (lecturers or speakers) and undergo training of between three- and six-months' duration.

The next institution selected was the Mauthausen Memorial Site, whose education program contrasts with that at Dachau in that a communicative focus is adopted when training staff (duration of training: 8 months). Referred to as *Vermittlerinnen* (f) or *Vermittler* (m) (mediators), guides at Mauthausen alternate between the three pillars of the institution's knowledge transmission concept—topographical orientation, historical clarification, and visitors' prior knowledge—which they must bring into balance in interaction with the visitors (Angerer et al. 2015, p. 11).

The perspectives acquired at these historical sites were contrasted with the educational approaches adopted by "neutral" locations. Thus, the research at Dachau and Mauthausen was followed by studies at the United States Holocaust Memorial Museum (USHMM) in Washington and the Anne Frank Zentrum in Berlin. At the USHMM, Holocaust survivors play a central role in the education work and help to "unlock places." These survivors act as VIP guides and guide high-ranking personalities around the museum. In addition, many retirees apply for training as a guide (duration: 13 weeks), after which they work as volunteers at the museum.

Like the USHMM, the Anne Frank Zentrum is not located at a historical site. However, the area on Rosenthaler Strasse in Berlin is replete with memories as it is adjacent to Otto Weidt's Workshop for the Blind, where Jewish people found work and a hiding place during World War II, and to the Silent Heroes Memorial Center, which commemorates unknown courageous Berliners who helped Jews escape Nazi persecution (Ballis 2017). One focus of the program at the Anne Frank Zentrum is on peer education. Guides are young adults who interact with the visiting school and youth groups on equal terms and help them to grasp the topic. The center succeeds in recruiting students aged between 18 and 25 years. The guides, who are called *Begleiterinnen* (f)/*Begleiter* (m) (peer guides), undergo a three-day training. At all four institutions studied, guides who have successfully completed their training have the possibility of participating in regular meetings for the purpose of further training and sharing ideas and experiences (see Appendix).

The data for the present study are drawn from interviews with guides and educational staff (interviews with guides: Dachau $n = 8$, Mauthausen $n = 5$, Washington $n = 4$, Berlin $n = 2$; interviews with educational staff: Dachau $n = 3$, Mauthausen $n = 3$, Washington $n = 5$, Berlin $n = 2$). In addition, at Dachau, a training course for guides who also worked as city tour guides in Munich was observed. Furthermore, at all four institutions, I conducted field studies and observed guided tours. In addition, training documents (scripts, structure of the training) were examined and used for the analysis.

Because of the intermeshing of data collection and data analysis, the following questions situated in the tension field between the guides' training and their job requirements emerged:

- What core activities are required and expected of guides during tours at memorial sites and Holocaust museums?
- What position do these guides occupy within their respective institutions?
- What optimization suggestions emerge for their situations on site?

3   The Arc of Work and Its Further Development as an Instrument of Analysis

As already mentioned, the present study is bound by the methodology of grounded theory. The purpose of this explorative methodology in qualitative social research is, first, to explain everyday phenomena and, second, to discover their theoretical content. Hence, the method focuses on social participants and their everyday routines. One characteristic feature of grounded theory is that it does not stop at methodological assumptions. Rather, in the framework of the research process, it provides tools for data analysis. The notion that data must be broken down analytically and de-contextualized is formative. To do this, a multistage coding process is used. The process of analysis, reflection, and writing culminates in the formation of a middle-range theory that emerges from the data and is thus related both to social reality and to "'all-inclusive' grand theories" (Glaser and Strauss 1967, p. 33).

In their reflections on the social organization of medical work in acute care hospitals, Anselm Strauss and his colleagues found further ways of analyzing and representing diverse work processes. They reconstructed work processes—in their terminology "projects" or "trajectories"—and sought to identify the sequentially and simultaneously performed constituent tasks. Strauss (1985) developed the concept "arc of work" to capture the totality of these tasks. Over the course of several years, this concept was empirically grounded in collected data (Feindt and Broszio 2008).

Strauss (1985) defined arc of work as follows:

"An arc for any given trajectory—or project—consists of the totality of tasks arrayed both sequentially and simultaneously along the course of the trajectory or project. At least some of the arc is planned for, designed, forseen; but almost inevitably there are unexpected contingencies which alter the tasks, the clusters of tasks, and much of the overall task organization" (Strauss 1985, p. 4).

Although Strauss did not systematize his deliberations, different elements of the arc of work can nonetheless be operationalized: Tasks are the smallest analytical unit; they involve the performance of discrete and observable work processes. Tasks are grouped into clusters of tasks that comprise several work steps. In addition, Strauss distinguished different types of work—for example, information work—that comprise tasks with a common denominator in terms of their work logic (Feindt and Broszio 2008). For German-language research, Fritz Schütze translated "arc of work" as "Arbeitsbogen" and undertook a critical examination of the concept (Schütze 1999, p. 340 f.).

In Germany, references to the arc of work can be found in mainly application-oriented studies on schools and professional practice (Bromberg 2012, p. 305). For example, analyses have been conducted of professional practice in the school context (Bräu 2002; Feindt and Broszio 2008), and three studies have been devoted to other occupational fields (Bromberg 2009: work in trade union organizations; Ackermann 2005: psychological counseling work; Thräne 2003: work activities of driving instructors). In the aforementioned studies, the data analyses were based on the underlying concept of the arc of work. With the exception of Bräu (2002), this concept was used to analyze interview data (Bromberg 2012, p. 316).

The present study on guides at memorial sites and Holocaust museums follows such a design: The arc of work is used for the analysis of the data sets (interviews, field notes, and scripts). Particular consideration is given here to the studies by Bräu (2002, p. 249) and Thräne (2003: 20f.), who operationalized the arc of work for empirical research using institutional, social, and evaluation components following Strauss (1991, p. 71–98). In this way, on the one hand, the complexity of the guides' job requirements is outlined, whereby the focus lies more on institutional action than on individual training (Seltrecht 2016, p. 70). On the other hand, the simultaneous work processes at these places are integrated into the analysis. On this basis, the core activities of the heterogeneous group of guides are developed from the data and interpreted against the backdrop of a remembrance culture in a state of flux.

4   Presentation of the Results

In the data material presented in what follows, the institutional, social, substantive, and evaluation components of the arc of work will be related to the training and activities of the guides and the work processes of the institutions.

4.1   Institutional Component

The term "institutional component" refers to all types of work that relate to the institution and the maintenance of work processes. Besides planning and information work, it also includes articulation and delegation activities: Different tasks or clusters of tasks are divided between the individual staff members (Bräu 2002, p. 249).

The institutions observed attach great importance to "planning." This is already evident at the beginning of the respective training courses, when the participants are invited to observe tours conducted by active guides. At the USHMM in Washington, these observation activities are guided, and checklists are made available to participants in order to direct their attention to different aspects of the respective guided tours. The behavior of visitors during the tours is observed, as are the reactions of the guides to crowded spaces, their behavior at the beginning and the end of the tours, their questioning techniques, and their time management. Substantively, the extent to which the exhibits selected by the guides fit the chosen narrative of the respective tours is examined in detail. Moreover, in a strategy-oriented approach, the observers are asked whether the guides encourage visitors to engage in critical thinking, to talk, and to contribute to discussions, and whether a connection develops between the visitors and the topic. Besides visitors' frequently asked questions, the USHMM *Tour Guide Training Manual* also features a list of "characteristics of a good tour" that includes the advice: "Use the exhibit to tell the story" (USHMM 2016). In addition, the manual includes self-tests to check knowledge, and possibilities for documenting impressions.

Peer observations are also a central element of training at other institutions. Conflicts may occur when established guides feel that the observers make it more difficult for them to do their job. To facilitate adequate planning, materials are made available. Besides relevant books and texts that cover the history, education program, and particularities of the institution, important documents and texts are compiled in a reader or made available on an online platform. The question of information is closely linked to the institutional component. There are both guidelines for tours and scripts with historical and pedagogical "information."

In addition, all institutions display regulations that inform the visitors of the activities that are allowed or prohibited at the respective institutions. The documents addressed to the guides feature little information about their rights and obligations. The Anne Frank Zentrum in Berlin is an exception in this regard. The center's tour guide training manual includes information on financial remuneration and a job description. It also features information on the establishment of an employee representative body: At a mandatory closed-door meeting, two persons are chosen from the group of freelance workers. These individuals act as contact persons in any disputes and communicate relevant news; they are integrated into the organizational processes of the institution and take part in selection interviews and closed-door meetings; and they also participate in the development of guidelines for training new cohorts (Anne Frank Zentrum 2016, p. 102 f.).

Concerning "articulation"—Who communicates with whom?—the guides at all the institutions studied are encouraged to conduct their tours interactively. At the Mauthausen Memorial Site, guides are advised to use historically correct wording and also to anticipate the language needs of the participants. The Anne Frank Zentrum favors peer communication, which, as the following excerpt from an interview with a peer guide illustrates, can sometimes lead to irritation on the part of the accompanying teachers:

"And it sometimes also happens that the teachers don't from a younger person don't accept it when they take a different position as it were. And then a conflict may develop, or it simply happens that the teacher just says: No, that's not how it was" (INT_AF_01_17; translated from the German).

In Washington, the trainers favor an empathetic, inspirational, and respectful style of speaking. At the Mauthausen Memorial Site, interaction is one of the main pillars of the education concept. Rather than being presented with a complete narrative, tour participants are asked questions that encourage them to engage with the topic, the landscape, and their own views. One interviewee, a guide who was also a member of the educational staff, saw a connection between the role of the question culture and quality assurance:

"And there are often, well it appears to me, it is often not clear how thoroughly we should really engage with the topic, how thoroughly, how well enough we are supposed to be prepared for it, for this mediation work or for this situation where we are there with the students. How well are we supposed to, how self-confidently should be we able to somehow deal with it? Or is it enough for us to ask questions and to hope that they talk to each other, and we are the ones who are the medium that keeps the dialog going" (INT_MAUT_03_17; translated from the German).

Diverse demands are evident in the above excerpt from the interview with the employee. These demands are further multiplied by the fact that all the institutions face challenges that are fueled by the increase in the number of visitor groups. Up to now, the principle "one tour for all" has applied at all these institutions. However, both the increase in the number of tourists from all over the world in the summer months and the necessity to cater for specific groups— for example, students with and without a migration background—are leading to changes in the communication situation. How guides should react to this is not always outlined in detail. Rather, they are still encouraged to develop their own narrative.

Regarding to the "distribution of tasks," the institutions are bound by a hierarchy: Permanent employees, and persons who are working on projects are clearly distinct from the guides—although all employees regularly conduct tours as part of their jobs. Nonetheless, the unanimous desire is that all staff members should feel part of a "community." The following excerpt from field notes begins and ends with a quotation from the director of the Memorial Site Dachau, who was addressing participants at the beginning of a training course:

> "'The most important thing first. We say 'du' to you [we use the informal *du* form of address].' He explains that there are two underlying reasons for this: First, the aim is that the group will grow closer together that way; second, the *du* is kept up on site in order to belong to the group of guides. 'Almost 500 people work here,' he said" (PROT_ DA_10_16; translated from the German).

The informal "du" is also linked to the fact that importance is attached to personal contacts at the memorial site. As one member of the education department at the Memorial Site Dachau put it: "We still have this idealism that we want to form kind of a family. So that's why this personal contact is important to us" (INT_ DA_12_16; translated from the German).

The image of the family was also invoked at the USHMM in Washington to characterize social togetherness:

> "Everything is really informal you don't have to set up a certain time. What I mean so the guide sees me and he had an issue with the group or he wants to speak, anytime he can send me an email, call me, not only me, but other colleagues, too. But also [...] just on the personal side, just relationship we have with our guides, we do actually, I think the museum as a whole, we look at ourselves as a museum family" (INT_USHMM_11_16).

The longer one observes the work processes at the institutions, the more obvious the differences between the staff members become. These differences manifest themselves in closeness to or distance from visitors. The "clusters of tasks" are clearly regulated and graded according to position: Guides meet visitors in the grounds of the memorial sites and in the exhibition area of the USHMM and the Anne Frank Zentrum. Although educational staff and employees with more permanent contracts also deal with guided tours—often for VIPs—they are entrusted mainly with conceptual work and the administration and organization of the guide pools and are increasingly moving away from the operational side of the guided tours. This distance from, or closeness to, visitors correlate with the type of employment relationship: On the one hand, there are a large number of freelance workers, who are paid by the hour; on the other hand, there are a small number of permanently employed educational employees. One trainer stressed that the system was justified from the institution's perspective:

> "When the first guides were trained here, together with the cooperation partners it was specified to a certain extent that the financing, the remuneration should be so good, be supporting; in the sense that it is a tough job here and some preparation is needed, but it shouldn't be such that it suffices as a sole income as it were; so we have, as a rule it is not the case that someone who works as a freelancer for us and our cooperation partners can make a living from it, um, and that was always important to us because after all there are full-time memorial site educators or providers in this line of work and there are those who implement it" (INT_DA_12_2016a; translated from the German).

4.2   Social Component

The "social component" also plays an important role at the institutions studied. Subsumed under this term are types of work that structure and support the social togetherness in the work process—for example, dealing with disturbances, engaging in trust work and contact work, and negotiating the division of labor (Bräu 2002, p. 249).

The institutions studied attach importance to "social togetherness." Notions of community or family are linked to expectations: Guides want to be recognized—also by the management—and to be valued for their commitment. Therefore, they see it as a problem when they receive too little recognition for their work. As a guide at one of the memorial sites put it:

> "To be quite honest, the memorial site is just not a good place [...] But what happens to the people there is unbelievable. So, the greatest personalities who start to work there are suddenly moral cowards after five years and don't dare to say another thing;

they've been killed. Emotionally. And it's the place, but also the people there. And it's the incompetence of the management staff. Absolutely. Who have no interest in things running well" (INT_11_2015; translated from the German).

These statements illustrate that guides need support on site, and that their social well-being must be attended to. Although trainers are usually aware of this, it appears that the guides at memorials are expected to take care of their own well-being: "The place does something to people. See to it that you're okay. Take care of yourselves!" (PROT_DA_10_2016). At the USHMM in Washington, by contrast, the guides reported that they were "well taken care of" (USHMM_Memo_11_2018). Because the guides at the USHMM are volunteers, they develop a special relationship with the institution—money plays no role in this regard (INT_USHMM_11_2016a).

Due to the specific atmosphere, disturbances and irritations frequently occur. "Dealing with disturbances" is thus an integral part of the training. Based on the collected data, it can be stated that these disturbances are addressed cognitively on a meta-level. Role-play, or a more action-oriented approach to the issue, is less common. A certain sense of helplessness takes hold when visitors have no interest in the place and the topic, when they express empathy with the perpetrators, or when the teachers accompanying school classes are forced into the role of justifier or corrector. It becomes apparent how much the disturbances depend on the way the guides see their work. In Washington, for example, the local schools are integrated into the museum's program. For some volunteers, who are retirees and have a specific idea of how tours should be, the evident lack of interest on the part of the students is a dual insult: to them and also to the victims of the Holocaust. As one guide explained:

"We had some student groups from a particular high school in Washington and they bothered some of the tour guides. I mean the tour guides were disturbed by them. And it was their behavior, they did not interact much and some of that was like you know, 'yeah well well' and that disturbed the guides a lot. […] A lot of people here who are guides are doing this because of a personal commitment to this museum and the Holocaust. And so the behavior of the group, they feel very personal about that, it, reflects how they did their tour. It demands respect for the Holocaust or lack of respect for the Holocaust and they did get very emotional about that" (INT_USHMM_11_2016a).

To create space for questions that arise after the training and for conflicts that result from changes, the guides are offered various forums at the institutions. Aimed at promoting "trusting cooperation," these forums take the form of meetings held at regular intervals—monthly or quarterly. They may have the nature of further training, while at the same serving as an opportunity for exchange and team building. Those guides want to engage with educational staff is generally viewed positively. However, this desire on the part of guides is viewed quite ambivalently by the employees on site:

> "And afterwards they [the guides] usually stay for a coffee. That is downstairs, of course, it's often like a beehive there. For the colleagues' downstairs, it's really diffi-cult, it's often just about tolerable when you're working against the clock and they all come and want to talk and tell stories, but they also bring a lot of experiences with them from the group, what wasn't good, what was particularly nice today, and that's unbelievably valuable. That's extremely valuable" (INT_MAUT_03_2017a; translated from the German).

Time to establish and nurture contacts appears to be a scarce resource. For exam-ple, cooperations and contacts with other institutions are rarely an integral part of the institution. The Anne Frank Zentrum is an exception in this regard, as it is the German partner organization of the Anne Frank House in Amsterdam, with whom it concluded a cooperation agreement in 1998 (Anne Frank Zentrum 2016, p. 96). This led to the institutional intermeshing of the institutions. For example, representatives of the Anne Frank House in Amsterdam are members of the board of the Anne Frank Zentrum in Berlin. They can thus exert direct influence on developments there and share their resources and expertise—that is, their networks, archives, and experiences (INT_AFZ_2_2017).

Substantively, contact work can be identified at two levels. On the one hand, the guides seek to engage with the visitors; on the other hand, it is their declared aim to "unlock" the place or its artifacts. The aim is to establish connections between the place and the persons. "Connection work" is the term used to describe this type of contact work. It captures the interplay of personal and spatial components.

The social component also includes the "negotiation of the division of labor." This procedure is hardly provided for at the institutions studied. After passing an examination, the guides become part of a guide pool and are called on for tours; they can indicate their time and date preferences. The institution defines the framework and underpins it with specifications and guidelines as well as training content.

4.3 Substantive Component

Having addressed the institutional and social components, I shall now focus on the substantive component, and thus on organization, interpretation, development, and implementation work (Bräu 2002, p. 249). For the institutions, the development of a tour narrative by the guides can be identified as the "substantive core" of the arc of work. This narrative is expected to be developed independently and adapted to the objectives of the respective institutions—this is not seen as a contradiction. The objectives vary depending on the institution and its history: At Dachau, the "Path of Remembrance"—the route from Dachau railway station to the Memorial Site Dachau, along which prisoners were marched to the camp by the SS—is frequently mentioned; at Mauthausen, the communicative aspect is emphasized; at the Anne Frank Zentrum, guides are expected to initiate processes of democratic self-empowerment; at the USHMM, the interpretation of the exhibit is the core element of a successful tour. The severity and consistency with which violations of substantive premises are sanctioned varies from institution to institution. The educational director at Mauthausen tries to avoid these problems by ensuring when choosing guides that they are willing to espouse the established concept:

"But we've learned, in so far as we now pay more attention when admitting the training course participants to how much the people are willing and able to engage with this interactive concept. That is of course no guarantee that it always works fantastically, but we see the difference between early trainings and late trainings" (INT_MAUT_12_2016; translated from the German).

Logically, "definition work" is of great importance. It refers, first, to the use of historically correct terminology. For many guides, having historically correct knowledge is essential and a key yardstick: "Yes, I believe that the greatest change for anyone who does their tour wholeheartedly is the growth in knowledge" (INT_DA_10_2015; translated from the German). Second, various definitions—for example, of the Holocaust—are presented and reconciled. In this way, a fit is achieved between concepts used in research and concepts implemented at the institutions (USHMM 2016).

How much room guides are given for "interpretation work" is an interesting question. At the USHMM in Washington, an essential objective of the training is that guides will become interpreters of the exhibit. They are encouraged to focus on certain things and to admit gaps in their knowledge. They are also encouraged not to let themselves be overwhelmed by historical facts. Addressing guides at a USHMM training course, a member of the educational staff stressed: "Look, if you're here because you want learn everything about the Holocaust, that's great, but that's not what this course is for" (INT_USHMM_11_2018b).

Because the exhibit in Washington is considered powerful, it is assumed that it will guide the guides. A further aspect of interpretation work can be found in the data from the Memorial Site Dachau, where thought was being given to generational change—not only with regard to the educational staff but also in relation to the role of survivors and their importance for the knowledge transmission work on site:

> "I think that it was always awful at an event where a survivor is there, the first thing he is told, at a time when there will soon be no more survivors, um, but it is now, it is now also a fact, yes. […] That opens up opportunities, but naturally it also involves losses, and as an opportunity I see that the already beginning freer way of dealing with German history possibly enables an attitude that is no longer aimed at repression and defense. I see as a loss the fact that nobody can answer the questions any more, which as you know still remain, so when you, when you forget to ask a question, or it didn't occur to you because you're too bounded by your own horizon, or it isn't even clear to you yet, there'll no longer be an opportunity to do so" (INT_DA_10_2015; translated from the German).

Guides assess the role of survivors in a differentiated way. However, an overview of the use of survivor testimonies during tours shows that there is little room for individualized recollection—personal testimonies are often dropped due to time constraints.

Change is also becoming apparent at the institutions. In Washington, Dachau, and Berlin, it was pointed out that the exhibits would be renewed in the foreseeable future; after all, they are, on average, 20 years old. Various plans are being made to take account of changing media habits. Guides who work freelance are only marginally involved in this "development work." However, the data collected from the interview sample of Dachau guides document that they make highly innovative suggestions that stem from their experience: When asked about possible changes at the Memorial Site Dachau, the guides made a wide range of suggestions. For example, they suggested the inclusion of the subcamps, which many international visitors would not get see but which were relevant for the concentration camp system. Further, they developed ideas about forms of remembrance to honor the memory of hitherto overlooked groups by erecting memorials at the site; they considered spatial changes as well as the local particularities and how to make them accessible using digital technology, for example, augmented reality and animations.

The guides are tasked with "implementation work" in line with the expectations of the institution, which primarily means that they must develop a narrative of their own. To ensure that this succeeds in accordance with the guidelines and objectives of the respective institutions, texts and materials that are deemed relevant are compiled in different media (for example, online platforms, readers, USB sticks) and made available to the guides. Materials are considered to play a key role in ensuring the quality of the training and keeping it up to date. In the following interview excerpt, a member of the educational staff at Dachau describes how the materials have changed:

> "The greatest change, first, it's actually the choice of material, we now have, have different literature packages than we had eight years ago. As far as the topics are concerned, things have, I believe, opened up a little, to some extent, about things like perpetrators, that played only a limited role before, or just one text was included, now it's a major topic" (INT_ DA_12_16a; translated from the German).

4.4   Evaluation Component

To ensure the quality of arcs of work, it is important that they be regularly adapted to the actual project or trajectory status by means of processes of analysis, assessment, and reflection (Bräu 2002, p. 349). Types of work that can be subsumed under the term "evaluation component" include supervision work, feedback work, evaluation work, error detection work, etc.

When guides start work at a memorial site or a Holocaust museum, they are encouraged to self-critically analyze their conduct ("analysis work"). One requirement with which guides all over the world are confronted is to work independently. Not all guides succeed in fulfilling this expectation and the requirements associated with it. One salaried educational employee at the Mauthausen Memorial Site described this as follows:

> "Many people simply need very clear guidelines and a lot of guidance, and a few people need only a little guidance, if you really see it that way. That is very difficult, you have to be prepared to independently develop a lot of things. And not everything that you independently develop is paid" (INT_MAUT_3_2017a; translated from the German).

What is striking in this interview excerpt is how often the word "independently" is used: It refers both to the training and to the development of a tour. In addition, the employee makes it clear that working independently is not associated with any financial entitlements.

I turn now to the aspect of "assessment work." At all the institutions studied, the respective training courses conclude with some form of examination. At the USHMM in Washington, trainees give friends and family a tour through the

museum. This tour is accompanied by museum employees. Afterwards the cer-
tificates are solemnly presented. One educational employee explained why this
format was chosen:

> "They come to the museum in order to graduate from the course, it is like a course
> you have to tour your family and friends and colleagues through the museum so our
> trainees have to bring in their you know their husband their wives and children and
> other colleagues, people that you know them personally in order to give a tour. And
> to us we think that's a good way to start off your touring […]. Because who knows
> you better than your family? And that's gonna be the hardest tour that you ever give
> for that's you have to do that in order to graduate" (USHMM_ INT_11_2016).

These thoughts reflect the notion of "family" and personal connection with the
institution that guides the views and the concept at the USHMM in Washington.
This connection between the goals of the institution and the chosen examination
format also becomes apparent at the other institutions. At Dachau, for example, the personal motivations of the guides are suppressed in favor of historical
facts. Hence, it is only logical that the examination there takes the form of a
"dry-run tour:" Together with a member of the educational team, the candidates
walk through the site and present the tour they have developed; the educational
employee is addressed as a visitor. Actual visitors or participants in the training course are not included (PROT_DA_11_2016). At Mauthausen, the training
course concludes with a "shadowed tour:" After a practice phase, the educational employees accompany the guides during a regular tour. The training course
ends with a follow-up discussion and the conclusion of a contractual agreement
(INT_MAUT_12_2016). At the Anne Frank Zentrum in Berlin, a gradual transition from shadowing to an independently conducted tour is favored; the guides
indicate when they feel ready to take on educational work.

For the most part, the assessment of the guides' work by the respective education departments ends on completion of training. Thereafter, the guides are
exposed to the judgments of the visitors; only occasionally are they shadowed
by educational employees. At some institutions, this reduction in the density
of guidance and assessment is partially compensated for through "reflection
and supervision work." During training, the shadowing of future colleagues
gives rise to a great need for discussion; after completing their training, guides
can voluntarily attend meetings and exchange ideas and experiences. With
regard to supervision, the USHMM in Washington offers guides the opportunity to discuss or address pressing issues with a psychologist once a year
(INT_USHMM_11_2018b). At Mauthausen, the guides consider the consequences of the job for the individual to be challenging and dangerous. In a

circular argument, one guide drew conclusions about the quality of the development work from his level of personal commitment. He vouched with his high level of personal commitment for the quality of his mediation work, although this commitment had a negative effect on his well-being, and ultimately on the quality of his work:

> "Through a lot of private commitment and time and naturally unpaid concept work, I simply performed. And that is something that, that eventually totally, how should I put it, that causes you to totally burn out, and that then of course somehow eventually has an impact on the active mediation work" (INT_MAUT_3_2017; translated from the German).

With regard to the culture of "feedback," further development through shadowing and feedback is an integral part of the guides' job at the Anne Frank Zentrum—even after training: The manual for the introductory seminar provides clear information in this regard that ranges from the timing of to criteria for shadowing and feedback (Anne Frank Zentrum 2016, p. 104 f.).

5 Interpretation and Discussion of the Results

The core activities of the guides that can be gleaned from the pooled results of the analyses are commitment and independence (Fig. 1). Guides have to be prepared to commit to the job or to the institution and to independently perform their tasks. A lot is expected of them, and what is expected of them cannot be offset with money. This insight is realized most consistently at the USHMM in Washington, where guides provide tours as volunteers. Even during their training, they have to independently develop content. In doing so, they are supported with selected material.

First, this ensures that content is relevant; second, it means that content is controlled. Certified guides independently decide how many further training courses they attend, whether they have a need for exchange or supervision, and how many resources they have for the development of further concepts. The fit between the tour and the visitors—for example, in terms of language register, question gestures, and the selection of the route or the objects—must also be achieved and implemented through experience. The USHMM expresses its desire that the guides should identify with the institution and its goals and, in part, initiates this identification.

Freelance guides occupy a key position within the memorial sites and museums: They are the point of contact with the increasing stream of visitors whom

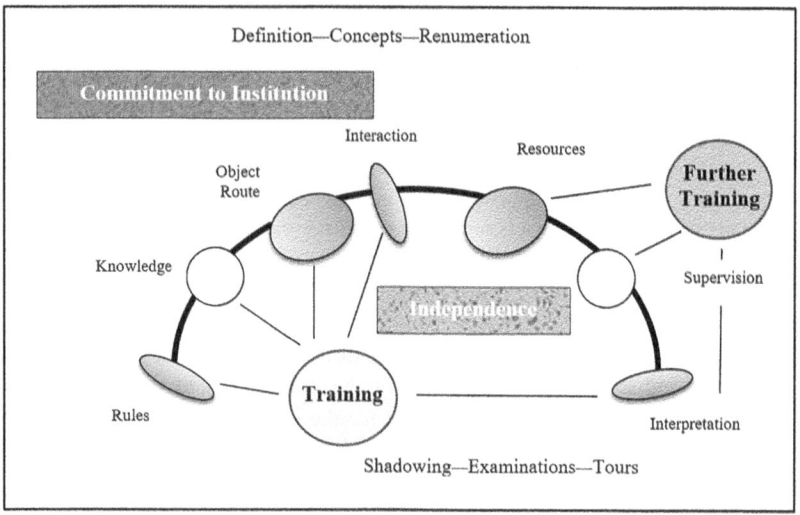

Fig. 1 "Arc of Work" for guides at memorial sites and Holocaust museums

they accompany through space and time. They "unlock" the place, set priorities during their tours, and act as representatives of their respective institutions. However, the analyses show that although they occupy a prominent position, they are rarely involved in the development of concepts and the selection of training content. At memorial sites and Holocaust museums, people with different interests meet. Only rarely do the institutions give guides the right to participate in the negotiation of rules and norms. As soon as they are trained and certified as guides, they move around the sites independently. The institutions allow them a degree of freedom to substantively implement their tours; further concessions with respect to institutional involvement are rarely specified.

Ways of optimizing the situations on site thus become more tangible: On the one hand, guides experience a deficiency in so far as they lack financial resources to educate themselves further and to carry out the challenging tasks. On the other hand, they perceive the job of guide as a resource that guarantees meaningful work and may offer possible promotion opportunities. It would be beneficial for all those working on site if rights and obligations as well as resources and support offerings were made explicit. It appears problematic when freedoms are suggested where commitment and voluntariness are expected.

At a time of far-reaching change in remembrance culture, the results of the present analysis are noteworthy in two respects. First, with their services, the guides make a direct contribution to the transmission of history: They engage in conversation with people, and—supported by objects and spaces—they can create access to the topics of the Holocaust and Nazi crimes. This should be integrated to a much greater extent into the guides' training, as opportunities to create such an access are rare in the digital age. It is certainly worth considering how guides might perform part of the task of the survivors in this regard and bear witness. Especially in light of the efforts of the memorial sites and Holocaust museums to digitally renew their offerings and exhibits, guides could acquire new tasks. They will not so much be replaced by digital offerings. Rather, they will be assigned new roles as mediators (see Ballis in this volume).

A first impression of these changes has been provided by a virtual reality (VR) testimony experience offered at the Illinois Holocaust Museum and Education Center. In a small, enclosed screening area—one person at a time—visitors view a documentary film entitled *The Last Goodbye*, in which Pinchas Gutter, a Holocaust survivor and former inmate of Madjanek concentration camp, revisits the site and tells his story. By wearing an audio-visual headset, visitors can experience the camp in 360° virtual reality. A guide helps the visitor to put on the headset and talks them through the forthcoming experience, mentioning, for example, that a former gas chamber can be entered virtually on the tour. The visitor is then taken on the journey to Majdanek with Pinchas Gutter. He guides them through the memorial site and recounts his experiences at the camp. At the end of their almost 20-min virtual tour, visitors often feel the need to talk about their experiences, and for this an approachable, listening, empathetic guide may be needed. Even under such changed conditions in the digital age, which are also emerging at other Holocaust museums and memorial sites, it remains a challenge to determine how greater harmonization can be achieved between the activities of the guides in their independence and the efforts of the respective institutions. Guides, too, need encouragement for their actions, assurance of their position, and appreciation for their often-voluntary commitment and the way that they process encounters.

It is a task for society to deter and combat xenophobic and antisemitic tendencies. The institutions should be aware of how strongly their objectives impact the mediation work of the guides. They should also be aware of the objectives that should shape the engagement with the Holocaust and Nazi crimes in the twenty-first century. With the instrument of the arc of work, the work processes at memorial sites and Holocaust museums have been made transparent. The results can help institutions and guides to reflect on the challenges of tour guiding in the twenty-first century.

Appendix

Overview—Guiding

United States Holocaust Memorial Museum, Washington D.C.	Memorial Site Dachau	Mauthausen Memorial Site	Anne Frank Zentrum, Berlin
Training course Focus: shadowing	**Training course** Focus: facts	**Training course** Focus: pedagogy	**Training course** Focus: peer-to-peer approach
Duration: 13 weeks	Duration: 3 or 6 months	Duration: 8 months	Duration: 2 days
Completion: Tour at the museum for family and friends; graduation ceremony	Completion: Exam in form of a dry-run tour; contract	Completion: Accompanied tour, certificate, contract	Completion: Undefined
Further training: Regular meetings, no quality control	Further training: Regular meetings, often themed	Further training: Regular meetings, 4 fixed-day meetings, quality control	Further training: Monthly meetings, shadowing, study trip, closed-door meetings
Some guides also work at other museums, where they also undergo training	Some guides also work at other institutions, some are also tourist guides	No other commitments	As a rule, students aged 18–25 years
Status Volunteers, no renumeration	**Status** Contract, paid per hour	**Status** Contract, paid per hour, travel expenses	**Status** Part of a network, paid per hour, and for further training
Job Title: Guide, not historian	Job Title: *Referentin* (f), *Referent* (m) (lecturer)	Job Title: *Vermittlerin* (f), *Vermittler* (m) (mediator)	Job Title: *Begleiterin* (f), *Begleiter* (m) (peer guide)
Dress code Red jacket, badge	Dress code Informal attire, badge	Dress code Informal attire, carrying a folder with materials	Dress code Informal attire, dressed like the visitors

United States Holocaust Memorial Museum, Washington D.C.	Memorial Site Dachau	Mauthausen Memorial Site	Anne Frank Zentrum, Berlin
Tour concept Human rights education Goal: connection Duration: 2 h	**Tour concept** Contextualization of history Goal: historical awareness Duration: 2.5–3 h	**Tour concept** Interaction and communication Goal: empowerment Duration: 2 h	**Tour concept** Learning from history for the present and the future Goal: social involvement Duration: 2 h

References

Ackermann, E. (2005). *Psychosoziale Beratung im Kontext pränataler Diagnostik. Möglichkeiten und Grenzen der professionellen Bearbeitung dilemmatischer Problemlagen.* Aachen: Shaker.

Angerer, C., Ecker, M., & Lapid, Y. (2015). *"Was hat das mit mir zu tun?" Zum Vermittlungskonzept der KZ-Gedenkstätten Mauthausen.* Mauthausen: BM.I. www.mauthausen-memorial.org/assets/uploads/paedagogisches-konzept.pdf. Accessed 12 May 2021.

Anne Frank Zentrum (2016). *Anne Frank. hier & heute. Handreichung zum Einführungstraining für die freien Mitarbeiterinnen und Mitarbeiter des Anne Frank Zentrums.* Berlin: Anne Frank Zentrum.

Ballis, A. (2017). Das Tagebuch der Anne Frank. Von analogen und digitalen Erkundungen zwischen Berlin und Amsterdam. *kjl&m, 3*, 31–42.

Bräu, K. (2002). Qualitative Schul- und Unterrichtsforschung. Zum Einsatz des Arbeitsbogenkonzeptes von Anselm Strauss als heuristisches Instrument zur Analyse von Schüler-Gruppenarbeit. *Zeitschrift für qualitative Bildungs-, Beratungs- und Sozialforschung, 3*(2), 241–261. http://nbn-resolving.de/urn:nbn:de:0168-ssoar-279635. Accessed 12 May 2021.

Bromberg, K. (2009). *Rekrutierung, Bindung, Zugehörigkeit. Eine biografieanalytische Studie zur sozialen Welt der Gewerkschaft.* Wiesbaden: Springer.

Bromberg, K. (2012). "Arc of Work"—als "sensitizing concept" für den Zusammenhang von beruflicher Arbeit und Organisationskulturen. In K. Schittenhelm (ed.), *Qualitative Bildungs- und Arbeitsmarktforschung. Grundlagen, Perspektiven, Methoden* (303–324). Berlin: Springer.

European Agency for Fundamental Rights (2011). *Discover the past for the future. The role of historical sites and museums in Holocaust education and human rights education in the EU.* Luxembourg: Publications Office of the European Union.

Feindt, A., & Broszio, A. (2008). Forschendes Lernen in der LehrerInnenbildung—Exemplarische Rekonstruktion eines Arbeitsbogens studentischer Forschung [55 paras.].

Forum Qualitative Sozialforschung/Forum: Qualitative Social Research, 9(1), Art. 55. https://doi.org/10.17169/fqs-9.1.314. Accessed 15 May 2021.

Glaser, B.G., & Strauss, A.L. (1967). The discovery of grounded theory. Strategies for qualitative research. Chicago: Aldine Publishing Co.

Gudehus, C. (2006). Dem Gedächtnis zuhören. Erzählungen über NS-Verbrechen und ihre Repräsentation in deutschen Gedenkstätten. Essen: Klartext.

Haug, V. (2015). Am "authentischen" Ort. Paradoxien der Gedenkstättenpädagogik. Berlin: Metropol.

Halbmayr, B., & Miklas, H. (2014). Die Perspektiven der Guides. In H. Bastel, & B. Halbmayr (eds.), Mauthausen im Unterricht. Ein Gedenkstättenbesuch und seine vielfältigen Herausforderungen (89–108). Wien: LIT.

Lautenbach-von Ostrowski, M. (2015). "Ich bin heute hier und ihr nicht mehr." Linguistische Fallanalyse von Zweck und Funktion einer Redewiedergabe in einer KZ-Gedenkstätten-Führung. In M. Becker, D. Bock, & H. Illig (eds.), Orte und Akteure im System der NS-Zwangslager (256–277). Berlin: Metropol.

Meseth, W. (2008). Schulisches und außerschulisches Lernen im Vergleich. Eine empirische Untersuchung über die Vermittlung der Geschichte des Nationalsozialismus im Unterricht. Journal für politische Bildung, 12(1), 74–83.

Österberg, O. (2017). Visits and study trips to Holocaust-related memorial sites and museums. In M. Eckmann, D. Stevick, & J. Ambrosewciz-Jacobs (eds.), Research in teaching and learning about the Holocaust. A dialogue beyond borders (247–272). Berlin: Metropol.

Schütze, F. (1999). Allgemeinste Aspekte und theoretische Grundkategorien des Werkes von Anselm Strauss für die Fallanalysen im Sozialwesen. In R. Kinsch, & F. Tennstedt (eds.), Engagement und Einmischung. Festschrift für Ingeborg Pressel zum Abschied vom Fachbereich Sozialwesen der Universität Gesamthochschule Kassel (321–347). Kassel: Gesamthochschul-Bibliothek.

Seltrecht, A. (2016). Arbeitsbogen. Reflexion professioneller Arbeit im Projekt- und Verlaufskurvenkontext. In M. Dick, W. Marotzki, & H. Mieg (eds.), Handbuch Professionsentwicklung (62–74). Bad Heilbrunn: Klinkhardt.

Strauss, A.L. (1985). Work and division of labor. The Sociological Quarterly, 26(1), 1–19.

Strauss, A.L. (1991). Creating sociological awareness. New Brunswick, NH: Transaction Books.

Thimm, B., Kößler, G., & Ulrich, S. (eds.). (2010). Verunsichernde Orte. Selbstverständnis und Weiterbildung in Gedenkstätten. Frankfurt a. M.: Brandes & Apsel.

Thräne, N. (2003). Professionelle Herausforderungen im FahrlehrerInnenberuf aus interaktionistischer Perspektive. Analyse der Problemstellen und Kernaktivitäten im Arbeitsalltag. In Perspektive. Analyse der Problemstellen und Kernaktivitäten im Arbeitsalltag. ZBBS—Zeitschrift für qualitative Bildungs-, Beratungs- und Sozialforschung, 4(2), 281–300.

United States Holocaust Memorial Museum (USHMM). (2016). Permanent exhibition. Tour guide training manual. Washington: USHMM.

Werker, B. (2016). Gedenkstättenpädagogik im Zeitalter der Globalisierung. Forschung, Konzepte, Angebote, Münster, New York: Waxmann.

Zumpe, H.E. (2012). Menschenrechtsbildung in der Gedenkstätte. Eine empirische Studie zur Bildungsarbeit in NS-Gedenkstätten. Schwalbach/Ts: Wochenschau.

Site Educators in Germany's Perceptions of Practice

The Sense-Maker and the Storyteller

Irene Ann Resenly

Abstract

This paper focuses on memorial sites as a place of learning in particular, it details the pedagogical perspectives of two educators at a German memorial site. Through indepth observations and interviews, the researcher explores site educators' roles in mediating Holocaust memory. Two questions are of central exploration: How do memorial site educators conceptualize Holocaust education? How do notions of history, memory, and place shape their ideas? Answering these questions through these two educators helps to shine a spotlight on the role of curiosity and context in their work through methods of sense-making and storytelling. Through the perspectives of these two site educators, the goal of this research to understand teaching as memory work and site educators as mediators of memory.

1   Introduction

Throughout Germany, former concentration camps have been preserved as memorial sites. Visitors interact with these sites in varied and complex ways. Some go to remember, others go to forget. Some visit the site by choice; others are obligated by family or schools (Clark 2011, p. 68–79; Weissman 2004; Zelizer, 1998). Thousands of German students visit memorial sites each year as part of

I. A. Resenly (✉)
PhD candidate, University of Wisconsin-Madison, Madison, USA
e-mail: resenly@wisc.edu

© The Author(s), under exclusive license to Springer Fachmedien Wiesbaden   33
Gmbh, part of Springer Nature 2022
A. Ballis (ed.), *Tour Guides at Memorial Sites and Holocaust Museums*,
Holocaust Education – Historisches Lernen – Menschenrechtsbildung,
https://doi.org/10.1007/978-3-658-35818-1_3

their public schooling experience. Holocaust memorial sites force them to confront an extreme manifestation of hatred and genocide. However, the sites do not exist in isolation. The staff of memorial sites shape visitors' experiences. Yet, we know surprisingly little about these educators and how they understand their roles.

This chapter focuses on memorial sites as a site of learning. In particular, it uses a case study (Yin 2009) as a methodology to understand how site educators perceive their role. It pays particular attention to their dynamic and purposeful pedagogical decisions. In this work, I seek to explore these questions: How do memorial site educators conceptualize Holocaust education? How do notions of history, memory, and place shape their ideas?

Site educators' critically engage with the history and memory of the Holocaust. As "mediators of memory," (Resenly, unpublished dissertation) site educators use the geography of the site to purposefully anchor their narrative, share complex survivor testimonies to illuminate Holocaust memory, and offer historical explanations that present diverse perspectives and often have no neat resolution. The process is deliberately provocative, as it prompts students to piece together the relevance of memory on their own.

Site educators negotiate complicated dynamics and tensions. They weave historical fact with memory and presentist considerations, centering meaning-making at the core of learning. This raises important questions related to the overarching ones, including: How do educators embody a social memory and re-embody a cultural and political memory of the Holocaust? What does this process of mediation look like? These educators have the dual position of being both representatives and producers of this entangled constructed memory of the Holocaust in Germany. This piece begins to address how that plays out in their teaching. For the two educators demonstrate how certain pedagogical values and methods can yield similar results. Let us meet the sense-maker and the storyteller by first understanding more about the context in which they teach.

2   Memorial Site Educators' Perceptions of Practice

In this paper, we encounter site educators' perspectives of their teaching through the lens of memory by offering rich descriptions of their practice. This was informed by observations of tours and pedagogical workshops over the course of 2015 and in-depth semi-structured interviews.

I wanted to understand the teaching of the Holocaust at a particular memorial site and how participants' everyday practices shaped teaching and learning at the site. In order to do so, I explored institutional knowledge; individual knowledge of the site educators, experience, and practice; historical and contemporary cultural contexts; and the theoretical terrain of Holocaust memory wrapped up in the teaching and learning of the Holocaust in Germany. Some of the work that influenced the framing of this research included the works of Assmann (2010), Bar-On et al. (1997), Clark (2011), Halbwachs (1992), Kohlstruck (1997), Marcuse (2001), and Young (1993). The site was selected for its complex history and for the willingness of the staff to participate in such a study. They were recommended to me by multiple scholars in the field of Holocaust education for their critical work in memorial site education and were well-suited and eager to engage in research that would contribute to their professional growth and development.[1]

This section introduces Wanda and Frank—two of the full-time site educators. It highlights curiosity as a guiding characteristic and value in both of their practices. However, curiosity manifested in different ways: Wanda's through sense-making and Frank's through storytelling.

Through the perspectives of these two site educators, the goal is to understand teaching as memory work and site educators as mediators of memory. I analyze how notions of history, memory, place, and pedagogy intertwine and intersect in their practice.

2.1   Wanda—The Sense-Maker

One memorial site educator I came to know was Wanda. She studied political science and worked at memorial sites and museums nearly her entire professional career. At the time of my research, she had worked at the memorial site for eight years. I observed Wanda lead eight school-group tours and facilitate four workshops.

When I asked Wanda how she would describe herself to someone who might not know her, she stated simply, "Curious and open" (Interview, 12/03/15). Curiosity permeated every aspect of her interaction with others. I witnessed Wanda converse with many people, and even the shyest, most reserved individuals related to Wanda well. Wanda was not only an engaging talker, but she was also a deep listener. She spoke with purpose and listened with intent. With

---

[1] However, due to the university's Institutional Review Board's policies for social science research, I cannot name the research site. The participants were given pseudonyms. I was granted the research opportunity under these terms in order to protect their identities.

little prompting, she was reflective on whatever someone might be talking about and could, in turn, ask questions of others.

Wanda's curious and open nature was characterized by playfulness and sarcasm, which Wanda readily admitted to with a smile. In everyday conversations, she liked playing the role of contrarian. This seemed to be the way in which she made others feel comfortable and willing to let their guards down so that they shared their perspectives. To sum up Wanda's curiosity in her own words:

"I'm interested in others. Motivations, madness, this whole colorfulness, what makes people—that interests me. That's the curiosity, and I am also an ironic person. I am very playful, and [so I like to try it out, asking] what is the opposite? How can I play with the opposite?" (Interview, 12/03/15).

By "opposite," Wanda was referring to different ideas or feelings that she might have experienced than from someone else.

When talking with Wanda, it was clear how invested she was in getting to know the person with whom she was speaking or exploring a topic of discussion. Her discourse when teaching had a strikingly meaningful pattern: speaking about something relevant to the person with whom she is engaging, bringing it to a bigger, current context or idea, and inserting humor and sarcasm into the conversation.

When I asked Wanda what was important to her as an educator, she stated simply "inspiration" (Interview, 12/03/15). When I asked what she meant by that, she said, "Maybe I'm just a catalyst between the history and the learner" (Interview, 12/03/15), adding that it was powerful for her to be fostering conversations at a place of perpetration. Wanda continued, rejecting the idea that acquiring historical knowledge was the only goal of her teaching practice and emphasizing inspiration:

"I am inspired; thinking inspires me. And nothing too sophisticated. I cannot expect something in their head immediately goes off and the inspirational words must come out. But it must be something, where I notice their circling thoughts. Maybe they [the thoughts] are not pronounced, rather there are hints that someone is formulating words, and I find that exciting" (Interview, 12/03/15).

Embodying this deeply relational act of teaching was at the core of Wanda's practice. She was passionate about sparking pedagogical moments that prompted students to work through difficult language in order to express a question or a response. She used the image of a wave as a metaphor to capture this process of her teaching: moving closer and further away, small and large, but, perhaps, always mighty. When observing Wanda on her tours, I recall with great clarity her being engrossed in and excited over a particularly powerful question a student posed or a fleeting moment when she noticed a change of expression on a student's face. Curiosity was more than just a personality trait for Wanda; it was a framework for her practice.

This process of questioning sense-making that Wanda valued and facilitated had different, but complementary, goals depending on if she was facilitating a tour or a workshop. With regard to a tour, Wanda shared that what mattered to her at the start of the tour was understanding the personality of the group of students. She asked questions to determine how to open them up if they were "intimidated," or make sense of a talkative group to see how she could organize their inquiries and interests. She termed that an act of mediation (Interview, 12/03/15). She continued, "It is really important for me in the first seconds to determine who is my audience, and what do they want? One has to say again and again, what are for me at the site the most important points that belong to their needs or might speak to [them]" (Interview, 12/03/15).

For Wanda, this approach could work in service of students feeling comfortable asking difficult questions at a memorial site, which she valued:

"And there can be fear and so they do not pay attention to the fact that they do not ask harsh questions at such a place [the site] because they think it is crossing the line or they cannot ask that [...] And there they have so many cultural-historical thoughts. And that's something, that's openness, so that's [them] almost ready to ask a question that would never ask at school, maybe even with their parents, here in this place" (Interview, 12/03/15).

In this context, Wanda was speaking of questions that might be considered "taboo" when she used the word "harsh." She spoke of the example of teenagers asking about menstruation at the site: How did female prisoners navigate that and how was it viewed by the perpetrators? Here she was inferring that there was tacit cultural understanding of what one was expected to talk about at a Holocaust memorial site. She viewed working against these cultural and gendered assumptions as a productive pedagogical challenge.

With regard to workshops, Wanda added the layers of connection between site educator, learner, and place. She said:

"I think about methods—that is to say I look at the story, on oneself, on the questions that should be asked. And that's what my goal is really. Hopefully, I can give them something through which they can express themselves" (Interview, 12/03/15).

Wanda's goal in a workshop was to support a sense of discovery at the site through methods that connected the story she was sharing and the learner's self-reflection and questions. She saw this discovery as tied to students' agency and interests. To Wanda giving a lecture was antithetical to discovery. She said: "'You must understand this history of the concentration camp.' That does not make sense to me. But I want to see what they discover." She wanted students to be active in their working with their own understanding and sense-making as tied to their experience at the place.

While communicating knowledge about the history of the place was not the focus of Wanda's practice (whether on a tour or participating in a workshop), grappling with students' reflections was deeply connected to memory and place. She stated:

"Maybe what is the point of historical place? So, the history took place 70 years ago. The place has changed. Why am I here? At a concentration camp, it is very important to play between the place, the victim, the perpetrator. That means, when I lead such a tour, I will always speak some sort of biography, always something about the individuals who have lived in the concentration camps—both those who have survived and those who have been murdered. So, it is also important to me to tell that this is the dead speaking about the dead. The place is not just a series of stones, and monuments and plaques—yes that has to do with the dead. It is a great challenge. And I do not want to name it a cemetery, because there are no sketched graves [...] but that sounds hard, practical, but for a family memory, it is not a grave. And that is why it is also important to sensitize this death" (Interview, 12/03/15).

For Wanda, interacting with the place was tied to engaging with the memories of the people who experienced it. She spoke of place as dynamic. Multidirectional aspects of memorial site learning entailed understanding the interactions with the place, victims, and perpetrators and how the place did something with the history of the site over time. Perhaps most vividly, learning at the site was about confronting state sanctioned murder. Moreover, in saying that for students the site was not a graveyard, Wanda was suggesting that the site was not a place of mourning but a place of confrontation and self-reflection. Wanda was working to foster sense-making among students by supporting students in (re)negotiating the meaning of learning about the place itself.

Finally, Wanda situated her work in a larger rhetorical context about the Holocaust in Germany. As she explained:

"Because there is a discourse in Germany, as if the memorials belonged to a kind of knowing. There is a great longing, and the word is always called normality. The creation of normality. The Holocaust, the Shoah, and National Socialism were the exceptional state of affairs, and now at some point [it] must be back to normal. If we hear something about Nazism, then someone goes to a memorial. They have achieved such a thing for a normality, but it is forgotten that this is the place of exception. For me [as well] still" (Interview, 12/03/15).

Wanda was talking about "normality" in two different ways: in the sense of collective historical memory as well as the purpose of memorial sites. The first "normality" was a nod to a perspective of German history in which the Holocaust specifically, and National Socialism more broadly, were understood as a break, an abnormality, in a historical narrative of progress and enlightenment (Marcuse 2001).

However, she also used "normality" to describe how the response to learning about National Socialism in the past, or confronting its modern-day iterations, was a visit to a memorial site. For Wanda there was an inherent tension in visiting a place of great tragedy. That tension was at the core of how she conceptualized the purpose of learning in that place. As curiosity and sarcasm distinguished her character, the tensions surrounding normality and irreverence led her pursuit of student reflection.

Wanda saw the role of teaching about the Holocaust as engaging with a "Bildungsort." This word connotes a sense of enlightenment, of moral exploration tied to place. Wanda's work as a mediator of memory was about prompting reflection in learners vis-à-vis place. She valued curiosity over content, questioning over answers, and breaking taboos over following timelines. For her, mediating the memory of the Holocaust was anchored in place and channeled through students.

2.2   Frank—The Storyteller

Frank always seemed to be reading something or talking about something he read. Whether we were on the train together or experiencing a rare moment of free time at the site, Frank had a newspaper or a book nearby. The reading material on Frank's desk seemed to be piled just a little higher than his colleagues'. Frank was an incredibly fast-talker with a lot to say, which was almost always connected to a question or an interest someone else expressed. His approach connected others' questions to something larger, and that united the speaker and him. Most often, he did so through stories. I observed Frank lead seventeen tours and facilitate five workshops for school groups.

Before I interviewed Frank, but after we had known each other for some time, I traveled to a different part of Germany and visited another memorial site. I was eager to share my impressions of this other site with those who worked at my research site, particularly regarding a sign laying out for visitors what counted as appropriate behavior. When I brought up the sign, Frank's curiosity was piqued, and he asked many questions about the sign—what it said, where it was located, and how people were interacting with it. He also shared his knowledge of that site including the history of the memorial and snippets of the surrounding town's reaction to it. He painted a complex picture even in our casual conversations. Frank introduced this idea of "appropriateness" into our talks and what it meant for students and teachers to grapple with, as opposed to implicitly accept, notions of appropriate behavior or respect at a site of tragic memory. Frank was a consummate storyteller. He combined a sophisticated use of images, complex narratives, and relevance to connect with students. I came to see this as a central approach to his teaching.

Frank had worked at the memorial for close to ten years; it was his first job. An internship tied to his studies in political science transformed into a job that became his career. When I asked Frank, what was important to him as an educator, he stated:

> "Mostly I think it is important that one continues to keep curiosity—curiosity about history, that one uses it to discover new facets, that new articles are continuously published permanently at [the site] or others. Of course, it's also a curiosity about people who come here with their questions. And simply to support that, as someone who helps them to take on this place with concrete problems and cases, and to then critically think about them. Yes, that also goes along with, through questioning, through sort of just addressing thinking processes and having some complete answers ready" (Interview, 11/15/15).

Like Wanda's, Frank's teaching practice was rooted in curiosity—supporting a curious disposition in the learner, a curiosity for history, and a curiosity towards the outcomes he was teaching. He enacted a kind of teaching that fostered learning as a process of discovery, and he was always ready to respond to learners. But while Wanda's curiosity manifested through the people she encountered at the site, Frank's curiosity seemed to show through the stories he told.

How Frank described himself mirrored what I saw in his practice. He was an encyclopedia of stories. At any moment, with seemingly any question posed, Frank could provide a historical story to share. Frank recognized his knowledge as being essential for his practice at the site:

> "I have already formulated some things, so that I am, so to speak, a kind of companion to the visitors who are here, and their questions, and am available, and also clearly want to convey the stories and also somehow ask others questions. If I don't

know, then I cannot ask questions. So, if I have no idea, that's hard for me" (Interview, 12/01/15).

His expertise in content, in the stories and history of the site, allowed him to connect to learners.

Moreover, Frank engaged with stories in three ways: by fostering historical imagination, offering complex historical narratives, and connecting students to history. Frank was the only educator I witnessed who consistently used images of historical artifacts to support the historical imagination of the students with whom he was working. Holding them up while he was speaking—sketches created by the people forcibly imprisoned at the site after the war or photographs taken by the perpetrators—Frank used the images to tell the story of the space. If a person stood at that same place on the site without an educator, they would hardly notice the small plaque that described what once stood there; the spot would easily be characterized as empty. Frank used the images to surface the stories of the place in the past and to support students in grappling with the site as it once was, as opposed to how it looked to them on the tour.

Frank also used stories to present a complex narrative of the history and memory of the site. For example, as he stood in the area where the stories took place, Frank recalled two survivors' testimonies that revolved around food and relationships. One was of a woman, who at the time she was imprisoned was deeply proud that her camp "sisters" trusted her to divide the bread rations equally. The second story had to do with a different young woman who saved tiny pieces of her daily bread ration for weeks so that she would have extra and might not go hungry on her twenty-first birthday. Day after day, she slipped another morsel under her pillow. On her birthday, not a crumb was to be found. Frank's stories never had neat endings. They always offered up complexity. There was no wrong or right, no black or white, no sense of redemption at the end. Moreover, it was the specific place where he stood that prompted the specific stories he told. The site offered an anchor to the narratives he connected.

Finally, Frank used stories to connect to students, responding to their questions and expanding their knowledge. A question about post-war justice inspired a story about survivors who spoke out at post-war trials. A question about liberation prompted a story about how someone was able to escape through the town, swap their prisoner uniform for some regular clothes hanging on a laundry line, and make their way to relative safety. A question about what townspeople knew solicited a story about the local butcher bringing food to the site and, according to the butcher's problematic retelling, allegedly knowing too late what was happening there. Frank had a meaningful story for virtually any comment or question that arose.

How Frank engaged with stories is part of a larger conversation about the work of history, memory, and pedagogy at memorial sites. In part, Frank saw his work as connected to narratives of remembrance and awareness. When I asked about his particular role in comparison to classroom teachers', Frank explained:

> "Already alone in that, I am active here as an educator and am constantly concerned with this topic, and also with students and our visitors negotiating this topic. I contribute, so to speak, with what would be remembered. That it is not forgotten. Of course, [the work of memorial site pedagogy] is important to me, too. Otherwise, I would not do this" (Interview, 12/01/15).

Frank was personally connected to this work and saw his role as an educator as active in constructing remembrance. Frank's insights elucidated how educator identity was shaped by an intentional, active formation of memory as part of practice. Teaching was memory work. As a storyteller, Frank's work (re)constructing memory was through stories- memories strung together to form cohesive, comprehensible moments or narratives. In teaching and learning at the site, he used these stories (in part), to awake awareness in students.

When Frank briefly discussed generational differences in learning about the Holocaust in Germany, he stated: "Well that's perfectly understandable when many students are here [at the site] and say, 'The history is already so far away for me.' For the first time there is an awareness. It's [the awareness] already there" (Interview, 12/01/15). Frank's use of the word awareness, or "Wahrnehmung," is significant because it infers that even though the history is distant, there is this almost innate knowledge of it by virtue of being German. Though the literature often uses the word "confront" to name the way in which Germans should engage with the history and memory of the Holocaust (Frei 1996), Frank's language suggested a process of surfacing instead. While "confront" means coming face-to-face with something challenging (often aggressively), "surfacing" infers bringing an idea to the forefront that might be inherently known. Surfacing is connected more to the notion of "Bildung" that must begin with an awareness.

In sum, Frank saw the role of teaching about the Holocaust as engaging with the history and memory of the site through stories. Frank's work as a mediator of memory was about building curiosity through stories of the site—his primary form of content of the site. He responded to questions that get at "so what" by elaborating on the "what"—offering up the context and content for students to potentially engage in the learning process on their own terms. Frank mediated the memory of the Holocaust through stories punctuated by place—not by providing answers, but rather offering information to students to reach their own understandings.

3 Significance of Wanda and Frank's Teaching

How each of the site educators sees themselves elucidates components of the work of memorial site pedagogy. Wanda's teaching revealed curiosity and its connection to sense-making as central to the work of memorial site education. Wanda's first act of mediation as an educator was between the group and herself. Next, she balanced the questions of the group with her expertise on what was valuable to learn about the site. The notion of inspiration was integral to her practice. Notably, Wanda was invested in using the site, using place, as a catalyst for addressing what might be taboo as opposed to using the symbolic power of the place to stun students into silence. By exploring what might be taboo at a memorial site, Wanda is pushing against what counts as being worthy or appropriate of being remembered. She is fostering complexity of learning at the site that challenges what might be typical of expectations of German memory around the Holocaust. By playing the role of contrarian, Wanda is also working to increase student engagement by offering more opportunities to respond, connect, or even push back. It opens up possibilities for students to assert their own thinking. She is challenging how one should engage at such a site—valuing confrontation and self-reflection over ritualized deference. In doing so, Wanda fosters a complexity of learning at the site that challenges the typical of expectations of German memory around the Holocaust.

Wanda saw her work as connecting stories, learners, and questions, while navigating the productive tensions of place, victims, and perpetrators in her pedagogy. She saw the work of learning at a memorial site as nested in German memory about the Holocaust and coming to terms with a feeling of normality both in regaining a sense of normal as Germans in the wake of the history and memory of the Holocaust and the idea of visiting a memorial site to combat the resurgence of a dark history in the present.

Frank's perspectives enriched an understanding of content expertise and its connection to the idea of curiosity through storytelling. For Frank, curiosity was rooted in the history of the site and in the students through the questions they posed. He supported this disposition of learning by having a robust knowledge base from which to draw critical questions. In the context of memorial site learning, he viewed his work as a content expert who can connect with learners by fostering historical imagination, offering complex narratives about the site, and connecting students to history. In relation to German collective memory around the Holocaust, Frank saw his role as actively continuing what will be remembered through work that matters personally for him. It was the stories that humanized this place of incredible dehumanization. The stories made empty buildings or open landscapes significant. The stories served as the content knowledge and, in turn, were how students would understand the narrative of the Holocaust. Frank prompted students to do something active with the stories, because he was always presenting multiple perspectives that were never quite resolved. Students would have to grapple or process to understand how those multiple perspectives made sense together.

4 Conclusion—The Role of Curiosity and Context

In essence, through curiosity and context and through sense-making and story-telling, Wanda and Frank are redefining what it means to engage with Holocaust memory at a memorial site. Their teaching practice is intentional, and their own reflections suggest their acute awareness of the role memory plays in guiding their practice. What Wanda and Frank's teaching shows is the possibility for questioning and exploration. Here, I do not mean questioning of the history of the Holocaust, but rather questioning the meaning of that history for the present.

Pedagogically, this shows us the power of curiosity and context in shaping German memory at Holocaust memorial sites. For, while their pedagogical processes have similarities, their differences demonstrate that they see no singular way for a learner to process the history of the Holocaust.

This marks a beginning of looking at the practice of educators at this site in particular and memorial site educators more broadly as mediators of memory. Though more research is needed to determine if their group of site educators are the exception or the new rule, their perspectives assist us in understanding their agency in memory work.

Site educators are often overlooked in the field of memory studies as playing a substantial role in how collective memory is constructed. They encounter thousands of students every year having arguably the farthest reach of intentionally shaping collective memory of the Holocaust for German youth. They themselves embody a collective, political, and social memory while facilitating that memory work for others. Therefore, we need to continue to learn more about site educators, as they are a key component to Germany's future memory of the past.

This case study centering Wanda and Frank's teaching shines a spotlight on the methods of site educators: sense-making and storytelling. They are both pushing against an assumed way to engage with the memory of the Holocaust. Because Holocaust memory is so entangled in German identity, Wanda and Frank—to some degree—are perhaps pushing against a right way to be German. For in the end, they are seeking to embolden German students to take ownership over how they take up the memory of the Holocaust.

References

Assmann, A. (2010). Re-framing memory. Between individual and collective forms of constructing the past. In K. Tilmans, F. van Vree, & J.M. Winter (eds.), *Performing the past: memory, history, and identity in modern Europe* (35–50). Amsterdam: Amsterdam University Press.

Bar-On, D., Brendler, K., & Hare, A. (1997). *"Da ist etwas kaputtgegangen an den Wurzeln ..." Identitätsformation deutscher und israelischer Jugendlicher im Schatten des Holocaust.* Frankfurt a. M.: Campus.

Clark, L.B. (2011). Never again and its discontents. *Performance Research*, 16(1), 68–79.

Frei, N. (1996). *Vergangenheitspolitik. Die Anfänge der Bundesrepublik und die NS-Vergangenheit.* München: Beck.

Halbwachs, M. (1992). *On collective memory.* Chicago: University of Chicago Press.

Kohlstruck, M. (1997). *Zwischen Erinnerung und Geschichte. Der Nationalsozialismus und die jungen Deutschen.* Berlin: Metropol.

Marcuse, H. (2001). *Legacies of Dachau. The Uses and Abuses of a Concentration Camp, 1933–2001.* Cambridge: Cambridge University Press.

Weissman, G. (2004). *Fantasies of witnessing: Postwar efforts to experience the Holocaust.* Ithaca: Cornell University Press.

Yin, R.K. (2009). *Case study research: Design and methods. Fourth edition.* New York: Sage.

Young, J.E. (1993). *The texture of memory: Holocaust memorials and meaning.* New Haven: Yale University Press.

Zelizer, B. (1998). *Remembering to forget: Holocaust memory through the camera's eye.* Chicago: University of Chicago Press.

Passing Down Testimony

How Concentration Camp Memorial Guides Mediate Testimony Through Linguistic Action

Moritz Lautenbach-von Ostrowski

Abstract

This article looks from a linguistic perspective at how a survivor of the Bergen-Belsen camp gives testimony and how his testimony is later passed on by a guide in the context of a guided tour. Three case studies show comparatively how a textualisation of oral testimonies is reflected in language and consolidated by third parties; how orality and textuality are interwoven in testimony; finally, how linguistic analyses are a useful tool for interdisciplinary Holocaust research and cultural-historical memory studies.

1 Guided Tours in the Prism of Linguistic Action

At memorial sites, guided tours for visitors are part of the educational practice, implementing two institutional purposes: Rememberance and commemoration and political-historical educational work (Heyl 2013; Garbe 2015). Although concentration camp memorial sites are attracting considerable media and political interest with increasing numbers of visitors and memorial education is establishing itself as a discipline in its own right (Gryglewski et al. 2015), there has been little research on guided tours at these sites. Jürgensen (2003, p. 55) noted already that an examination of the formalities of a tour would be relevant for guides, because they moderate the tour and are responsible for the instituons' goal formation, direction, and their own evaluation. While Gudehus (2006), Pampel (2007) and Haug (2015) look at guided tours primarily from an educational

M. L. Ostrowski (✉)
Universität Hamburg, Hamburg, Germany

© The Author(s), under exclusive license to Springer Fachmedien Wiesbaden   47
GmbH, part of Springer Nature 2022
A. Ballis (ed.), *Tour Guides at Memorial Sites and Holocaust Museums,*
Holocaust Education – Historisches Lernen – Menschenrechtsbildung,
https://doi.org/10.1007/978-3-658-35818-1_4

perspective, Ballis (2019) presents an ethnographically based study that focuses on guides and their working conditions.

To these works my contribution adds a linguistic perspective: In my Ph.D. thesis on linguistic action as "approaches to memory," I investigated forms and functions of linguistic actions during guided tours with a special focus on guides (Lautenbach-von Ostrowski forthcoming a). My work is based on empirical linguistic data of spoken language I grounded from guided tours and analyzed according to *Functional Pragmatics*, a linguistic theory and discourse analytical method. The corpus gathers six guided tours for schoolchildren aged 16 on average, I surveyed between 2010 and 2015 at the concentration camp memorial sites of Bergen-Belsen, Neuengamme, and Ravensbrück.

Despite its statistically small size the corpus contains a wide range of language material since it assembles 12 h of spoken language and about 120 participants. On the basis of 32 exemplary and qualitative case studies from four of these tours linguistic actions are examined, with which the tourguides convey knowledge and (historical) conceptions to their hearers: How do the *actants* of a tour (*speaker* and *hearers*) process and structure their historical knowledge through linguistic action? What are specific "gnoseological qualities" (Ehlich 1997) of these discourses? Are witnessing and testimony conveyed during these tours and if so, by means of which linguistic-mental forms? The results of the linguistic analyses are being discussed with regard to memory studies, didactics, and historical science. One of my central findings is that throughout their tours guides do convey historical, often site-specific knowledge, which they transfer in a structure that can be grasped as *image-knowledge* (Ehlich and Rehbein 1977). This offers a link to educational research in the field of history, which has shown further interest in mediation concepts (Zülsdorf-Kersting 2007; Lange 2011).

In the course of my work, I had the opportunity to present aspects of linguistic action in guided tours. The results of these articles are only briefly outlind here: In Lautenbach-von Ostrowski (2014), I introduced a conception of linguistic "approaches to memory" considering the role of actional and linguistic pointing at the historical sites (deixis). I further focused on the significance of the reproduction of survivors' speech by guides during their tours, as well as metaphors and comparisons (Lautenbach-von Ostrowski 2015, 2020); finally, I looked at forms of narrating during tourguiding, what characterises a guide's narrative and which other linguistic forms are also (even more) relevant (Lautenbach-von Ostrowski forthcoming a). forthcoming b!

2   Functional Pragmatics and Guided Tours

Functional Pragmatics grasps language as a socially and historically developed tool and a purpose-bound form of human action.[1] Partly following Austin's speech act theory and Karl Bühlers notion of Deixis, Functional Pragmatics provide a consistent theory of linguistic action (Redder 2008), mainly based on empirical studies of spoken language, especially in the field of communication at institutions. The analyses focus on the purposes of linguistic action and reconstruct how speaker and hearer act with each other depending on constellation and speech situation. In this context, purpose and function are central analytical categories: Language has three basic functions, working in a teleological, communitarian and gnoseological manner (Ehlich 1997). Linguistic action aims at the processing of (non-linguistic) reality and knowledge. It is carried out in linguistic action patterns (for example question and answer), from which individual linguistic action is derived. A linguistic action is subdivided into three sub-acts or dimensions: the phonetic act of utterance, the intellectual proposition and the action dimension (illocution). People have a mental realm (Π, Pi) in which reality (P) is reflected to them and which contains their knowledge, ideas, fantasies and expectations. Speaker and hearer access their knowledge (mental action) in their linguistic actions (p). Functional Pragmatics focuses on this relationship between thinking, speaking, and reality (Ehlich and Rehbein 1986) including *knowledge*; actants can structure their knowledge in a specific way by means of linguistic action. Speaker and hearer meet in *constellations*; the term grasps a situation including its action-relevant moments (Rehbein 1977, p. 265). Within institutions speaker and hearer meet as *agent* and *client*. Institutions are socially and historically developed action apparatuses that serve social purposes (Ehlich and Rehbein 1986). Between agent and client there is a systematic knowledge gap.

For certain institutional constellations, *linguistic action patterns* and forms of action have developed that serve to overcome such knowledge gaps. A central component of a constellation is the speaking situation, in which speaker and hearer interact with each other. A fundamental distinction is made between *text* and *discourse*. Discourse implies the co-presence of speaker and hearer (tendency:

---

[1] Linguistic action is interactional action, other forms are actional and mental action. Alternative names for the basic functions are: communicative-purpose-related, community-related, knowledge-related. As a theory and discourse analytical method Functional Pragmatics is mainly developed in German language. Central terms used in this contribution are italicised, e.g. *speaker* and *hearer*. Also, the empirical data presented here refer to a guided tour held in German. Thus, the translations of scientific terms as well as of the empiric linguistic data given here are tentative.

orality) whereas text culture-historically derives from discourse, closely linked to the genesis of institutions. Text does not have to appear in writing, but above all it implements other purposes than discourse: it overcomes diatopia and diachrony, place and time, and thus stretched speech situations between author and reader (analogous to speaker and hearer). Text aims at the transmission of language including its action dimension (Ehlich 1983). Text and discourse are also defined as the most complex units of linguistic action.

A frequent phenomenon in my corpus are forms of survivors' speech reproductions given to the hearers by the guides. From this empirical observation I derived the assumption that this has to do with the gnoseological potential and functionality of speech reproductions: With their help a speaker can let a hearer participate in past constellations and actions and provide insights into the knowledge and emotions of historical actants ("testimonial function," Brünner 1991). Functional-pragmatic studies on the quality of action in (fictional) literature show that this is also effective if neither speaker nor hearer were present at the original event and that it also applies to the quality of action of made-up speech renditions (Redder 2003; Lautenbach-von Ostrowski 2020, forthcoming a). Further, in my studies I found that guides often reproduce all three dimensions of a speech action, which is the utterance act (phonetic), the propositional content (for which guides sometimes use alternate wording), and the illocutionary act, i.e. the action quality of a speech action.

In light of this theoretical background I made use of the primarily cultural-scientific concepts of cultural and communicative memory for linguistic theory and analysis: Aleida Assmann (2006, p. 208) assumes that visitors at concentration camp memorial sites could experience "initiations into a collective memory" and "transitions from the past to the present." My study shows that and how these initiations and transitions are linked to linguistic action. With regards to the three basic functions of language (Ehlich 1997), Assmann's initiation is to be understood as a communitarian aspect of guided tours insofar as they aim to create a community of memory. Reasoning and argumentations as to why there are concentration camp memorials are a teleological aspect of guided tours.[2] A gnoseological, knowledge-related function of guided tours aims at overcoming distance in the knowledge of the hearers. In this sense, the three basic functions of language are working within in the guided tours. They appear to be communicative aims which the guides fulfill in their practice by the use of linguistic actions I refer to as "approaches to memory:"

---

[2] Reasoning and arguing are illocutionary connections that can extend over longer chains or sequences of linguistic actions.

(A) Pointing: Using deictic expressions a speaker can point to somebody, some place or time. "I," "you," "we," "here," "there" are expressions, for example, that work within reference spaces of linguistic action (space of perception, space of speech, text and knowledge; Ehlich 1982). Linguistic pointing can replace or support actional pointing and is possible on a spatial as on a temporal level. It is also particularly dependent on speaker, hearer and location. Thus with deictic expressions guides can orientate their hearers in the terrain of historical sites, but also point within the cognition and ideas of their hearers.

(B) Reproductions of speech, forms of reported speech: These forms make historical actants and their actions, emotions, etc. present to current hearers. They are also suitable linguistic means to illustrate knowledge, to convey experiential knowledge and to involve current hearers in contexts of action in which they were not themselves present (Brünner 1991). These qualities make forms of reported speech useful means to re-oralize history within guided tours.

(C) Metaphors and comparisons: They can also serve as illustrations. In particular, they serve to transfer knowledge from a known field to an unknown one (Lakoff and Johnson 1980) and act as means of building and expanding knowledge in expert-layman-communication.

In this perspective guides at concentration camp memorial sites act as memory specialists (Jan Assmann 1992, p. 54). Linguistically speaking, they are institutional agents who function between cultural and communicative memory by processing knowledge from one holding to the other: They liquefy historical knowledge frozen in cultural memory by bringing it back into discourse and making history with its experiential dimension present to their hearers. In doing so they can also function as witnesses of the eyewitnesses or *messengers* (Ehlich 1983) of the survivors. A cultural-historical precursor of written texts, the messenger was obliged to transmit linguistic actions, including their propositional and illocutional dimensions.

As analyses in the above-mentioned corpus show, due to their testimonial function speech renditions are suitable for the guides to portray or illustrate historical constellations, propositions, ideologies etc. (Lautenbach-von Ostrowski 2020, p. 22–30). Even though guides often reproduce the propositional content and illocution of an original utterance quite accurately while editing its phonetic level (Lautenbach-von Ostrowski 2014), this also applies to 'freely invented' speech renditions. Other examples show how close, even identical wordings are, with which guides reproduce the speech of survivors and convey the dimensions of their linguistic actions (illocutions). On the one hand, this reveals the startling relationship between textual, cultural memory and discursive, communicative memory: From a linguistic point of view cultural memory tends towards text, communicative memory tends towards discourse; guided tours at concentration camp memorial sites are constellations positioned in between. On the other hand, it becomes obvious to what extent guides convey testimony during their tours. The following case study shows this relation by analyzing how one guide reproduces the speech of a survivor quite precisely on all levels of linguistic action.

3   Moshe Nordheim's Testimony

The focus is on excerpts from three versions of Moshe Nordheim's testimony. Concerning their propositional, illocutionary and utterance dimensions they are related to each other. Thus, the corresponding transcriptions are presented below in longer versions: Archive is taken from a biographical interview Nordheim gave for the Bergen-Belsen memorial archive in 2000; NDR is taken from an interview he gave to the German broadcaster NDR in 2015; Tour documents how a guide at Bergen-Belsen passed down Nordheim's account to her hearers in 2012.

Moshe Nordheim was born in Amsterdam on 28 January 1934. The Family (parents and younger siblings) was persecuted by the National Socialists for being jewish. In 1944, when Nordheim was barely ten years old, they were deported to Bergen-Belsen. He writes about himself on the website of the concentration camp memorial:

"I was a prisoner in Bergen Belsen from 12/1/1944 till 10/4/1945. Then I was taken from there by the 'lost train' and was liberated on 23/4/1945 in Trobitz by the Russians. I wrote a book [...] *From Rebuke to Rejoicing*. In the Stiftung you can find a very long interview on a video [...] in which I tell my story. This video is used [...] for pedagogic purposes. I gave in the past many lectures in the Stiftung and in schools in the neighbourhood of the Stiftung. [...] I have, after 70 years, very bad and dark thoughts. Because my feelings are that the world has nothing learned from what happened in the holocaust. [Antisemitism] grows, and Jews are not safe in Europe. And then I ask myself, have we today a come-back to the dark years 1930 etc. [...] It is very important to educate the next generations. Therefore I wrote a book, I give lectures etc. which will remain after my death" (https://fsjgbb.wordpress.com/umfrage/, corrections made by the author of this article).

Nordheim has told his story many times throughout many years and in different constellations. While dealing with the guide's rendition of Nordheim's testimony in my research process (Tour), I came across his interview from 2015 (NDR) and realised that he himself depicts an event to which the guide refers. Because of the similarity of these accounts, I also looked at Nordheim's biographical interview from 2000 (Archive), which is about five hours long and in which he also portrays the event. The comparison of all three versions shows that both of Nordheim's depictions, between which there are after all 15 years, and the guide's accounts of his experiences are very similar. In the following the focus is on a childhood-memory of Nordheim when he had to stand for roll call in the so-called Sternlager of Bergen-Belsen and got into a potentially life-threatening situation for camp inmates because he needed a toilette urgently. What *pieces of knowledge* does he verbalise repeatedly and how are they passed on by the guide as she takes on his mission to "educate the next generation"?

4   Transcriptions and Analyses

4.1   Archive (2000)

The excerpt is taken from the biographical interview with Nordheim from 2000, which is in the archives of the Bergen-Belsen Memorial and known to the guide at the time of the tour in 2012. At this point, the interview has been running for about three hours. Nordheim tells about the episode on the roll call square, which is particularly considered in the analysis; propositionally, he still refers to it afterwards (score areas 9–18; score areas short: sa.).[3]

The transcription Archive (Appendix 1.) documents a completely different speech situation than the one called Tour; the interviewer in Archive and the guide in Tour are not the same person. The transcription begins at a point where the interviewer asks about Nordheim's specific experience of the roll call in the concentration camp as a child: "How was it when you had to stand there for hours as a child, that's something else than as an adult?" (Archive, sa. 1 et sqq.). A comparison with the transcript NDR shows that Nordheim is asked a similar question: "tell me again please a little bit more precisely, what did it mean to be in a concentration camp as a ten-year-old child?" (NDR, sa. 3–5). In both cases, the interviewers act empathetically and cooperatively by anticipating the

---

[3] The transcriptions are according to the HIAT convention using the EXMARaLDA Editor (Ehlich and Rehbein 1976; Rehbein et al. 2004).

situation of the witness: In Archive this shows by the interpolated assertion of the interviewer "as an adult, you know you have to do it now, but children are often/ they don't want to or they have to" (sa. 2 et sqq.); in NDR this shows, for example, by the fact that the interviewer lets Nordheim determine his exact age at the time (NDR, sa. 1 et sqq.).

In both speech situations Nordheim is confronted with similar questions, both interviewers emphasise the special moment of childhood in the camp and ask the witness for information about it, but also leave it up to him what he wants to say about it. Against this background it can be assumed that Nordheim is familiar with questions of this kind and has developed strategies for dealing with them, and this can be seen in his answers: with questions concerning the connection childhood—camp—roll call and the memories they evoke, Nordheim obviously copes by verbalising three interconnected elements of his biographical knowledge, that for him have an exemplary status. "I can tell you three stories" (Archive, sa. 5 et sqq.): (A) the game of warming hands, a depiction in which he also performs gestural repetitions (Archive sa. 6–10; NDR, sa. 12–16); (B) the "episode" of the urge to urinate during roll call, in which he repeatedly and similarly makes use of reported speech (Archive sa. 9–14; NDR, sa. 18–22); and (C) his reference to the particularly poor condition of the clothes (Archive sa. 19–25; NDR sa. 29–35). Also striking is a structural similarity of the witness' answers: in both cases he says that he will "tell three stories" or has "two things to say" (Archive sa. 5 et sqq.; NDR sa. 11 et sqq.). He then even arranges pieces of his knowledge in the same succession: first, hand game; second, roll call; third, the clothes.

4.2 NDR (2015)

The interview referred to here as NDR is no longer freely accessible on the internet, but is in the regional broadcasting centre's archives in Hannover. The online version was about 55 min long, from which I transcribed approx. 4 min.

This transcription NDR (Appendix 2.) shows that Nordheim is asked to "tell a little more precisely" what it meant to be a ten-year-old child in a concentration camp (sa. 1–6). Accordingly, he elaborates in more detail on his possibilities for action at the time and the mental dimensions of his childhood under the conditions of existence in the camp in 1944. He assesses his childhood and looks at it from the result (sa. 6–10). Apparently, he mentally worked out the loss of his childhood with a certain acceptance: "I know that! Just what can I do?" (sa. 10). Then, in a similar way as in Archive (sa. 5 et sqq.) Nordheim prepares the passing down of pieces of his knowledge—propositional cores of memories he beholds as important for his testimony: "And then I have two things to say" (NDR; sa. 11).

4.3 Comparing Nordheim's Representations

Looking at the transcriptions Nordheim's depictions are quite similar to each other concerning the three dimensions of his linguistic actions (utterances, propostions and illocutions), the order of the historic constellations he presents and how he worked on them mentally. This is also remarkable since there are 15 years between the interviews. Regarding the structure and order of his depictions in both cases Nordheim presents the warming of the hands, the urge to urinate during roll call and the state of the clothes in the same way. These elements of knowledge appear to be particularly relevant components of his testimony; they exemplify his childhood experience in the camp and illustrate his possibilities for action at the time.

Regarding the conclusions of the three episodes, their evaluation, and the transitions between them they are also similar. With the hand game, Nordheim connects the punch line that "children always find something to do" or can be especially resilient: "Even though I was an adult, we were also children, we wanted to have warm hands!" (Archive, sa. 4; NDR sa. 15 et sqq.). Regarding the roll call, Nordheim emphasises the brutality of the situation (Archive sa. 10–26; NDR sa. 28) as well as his mother's and the other prisoners' fear. He speaks outright indignantly about the condition of the clothing, both times referring to it as "cardboard" (Archive, sa. 20; NDR sa. 34), once using the Yiddish expression "Schmáte," for rags (NDR; sa. 31). These rags made it considerably more difficult for prisoners to survive, especially under harsh weather conditions (cold, rain; see transcriptions), an experience emphasised in many survivors' testimonies and also conveyed by the Guide (Bock 2017, p. 116–153). Also the depictions of the roll call situation resemble each other in their dramaturgy, including conflict, tension and resolution (e.g. "And that was long!," Archive, sa. 16 et sqq.; NDR sa. 22). Particularly striking are the almost identical speech renditions with which Nordheim puts the historical constellation 'into scene' (Archive, sa. 12 et sqq.; NDR, sa. 18 et sqq.).

Although his depictions of the roll call are similar in their basic assessment ("terrible;" Archive sa. 18 et sqq., NDR sa. 28 et sqq.), in the NDR interview Nordheim shows more understanding towards his mother and their fellow prisoners and links the three episodes more explicitly to each other: "And that you have to take everything into account that you live no longer in the accustomed world" (NDR, sa. 34 et sqq.).

Finally, Nordheim explains to his (potential) hearers that there was no security or reliability for the prisoners of Bergen-Belsen. The camp was no part of the accustomed world. From the perspective of linguistic action, the witness thus shatters knowledge on which common bases for action are founded in the form of expectations. These kinds of assumptions regarding normality are anchored within the cultural apparatus (Redder and Rehbein 1987). Such ruptures of knowledge are typical for Holocaust testimonies (Bock 2017) as well as for guided tours at concentration camp memorial sites (Lautenbach-von Ostrowski forthcoming a, b).

4.4   Tour—The Use of Nordheim's Testimony in a Guided Tour

The transcription in Appendix 3 shows how a guide in Bergen-Belsen reproduces Nordheim's testimony on a tour for high school students. While she puts the focus on the roll call situation she does not mention the hand game. Still, the comparison shows that she also conveys the propositions and illocutions of further utterances of the witness. For example, she stresses the importance of clothing for the victims.

At the time documented in Tour the group stands at the site of the former roll call square of the so-called "star camp." Since the square is no longer visible, the guide first points out to the hearers where they are now. She does this by using deictic expressions (Appendix 3. *italicised* here), pointing onto the group in their shared space of perception: "*We* are *here now*." In this way, the guide makes the historical place imaginable to the hearers: "At this place, in camp time, that is when there was (a) concentration camp *here* [...] *here* was the roll call site of the Star Camp" (Tour sa. 1–3). Linked to her linguistic pointing within the actant's shared *space of perception*, the guide explains the historical function of the square and additionally evaluates it, with the aim of expanding the hearers' knowledge ("people had to line up," "ususally lasted two to three hours," "without regard whether it was a child or an adult," "it was counted," "mere chicanery;" Tour sa. 1–6).

With reference to the site as it is today, the guide outlines its historical appearance; she does this within the hearers' imagination (and knowledge): "[...] when there was a concentration camp here" (sa. 2 et sqq.). Only after informing the hearers about this place and what used to happen there, the guide passes down Nordheim's testimony—the exemplary narrative (Bredel 1999, p. 59–70) that

Nordheim himself formulated. As to how the guide does so, three aspects are relevant: First, she keeps two of her speech reproductions particularly close to Nordheim's utterances. Regarding the dimensions of a speech action she adopts propositional contents and illocutions; concerning the act of utterance, she even adopts the child's suffering tone of voice (sa. 9–14). Second, with her third reported speech the guide summarises further knowledge and propositions of Nordheim without quoting him exactly. For instance she takes over his punch line of relief and turns it into a question for the present hearers: "[…] and no one had seen it. Important. Why?" (sa. 14 et sqq.). This utterance aims at mental actions of the hearers to ponder and conclude why the depicted situation was extremely dangerous for camp inmates. In this sense, the anecdote opens up a perspective from the particular to the general, from the example of one survivor to the fate of hundreds of thousands of victims. Third, with regards to the concept of the cultural apparatus, this also has consequences for the hearers: being confronted with testimonies of survivors of the camps can sensitively disturb their knowledge and expectations regarding normality.

4.5   Outlook and Conclusion

At first, my analyses focus on two speech situations and their references to each other: two versions of Moshe Nordheim's testimony show to what extent a survivor's testimony solidifies or changes over the course of 15 years. Then, a case study of a guided tour at the Bergen-Belsen memorial site illustrates how a guide passes down Nordheim's account: she has recognised which pieces of knowledge are relevant for the witness and adopts them in her speech, in which she includes propositions and illocutions, in part even exact wordings of the witness. In this respect she acts as Nordheim's messenger. This perspective shows how linguistic analysis can underpin cultural scientific theories of memory: cultural memory can be grasped as *text*, communicative memory as *discourse*. As social memories they are interwoven on linguistic and mental levels. The original *speech situation* in which Nordheim bears witness is *discourse*. His testimony is archived and becomes part of cultural memory i.e. *text*. During the tour the guide then performs a return of this text into the discourse of the communicative memory of the group. The analyses reconstruct the path that the witness's knowledge takes in this process. It becomes apparent how the guide both takes on the mission of the witness to "educate the next generation" and linguistically implements the educational aims of her institution. In this respect she acts as a memory specialist who mediates between social memories.

In Holocaust research, the concept of the gap is well known, which testimonies of survivors of the camps always contain (Agamben 2002; Levi 1986). From an oral history perspective Kabalek (2015) stresses that survivors share their memories in very best versions that contain peaks and essences. In this sense, pointedly speaking, Kabalek also observes a textualisation of testimonies. As a whole, the testimony of Moshe Nordheim presented here may also contain such gaps and the transmissions discussed here may also be very best versions of the professional narrator Nordheim, in which the hand game, the experience on the roll call square and the memories of the clothes form peaks or essences.

Linguistic analyses can be a tool for identifying such aspects of knowledge and for tracing how they are passed on via linguistic action and what mental processing they undergo on the part of the speaker and possibly also the hearer. In this way, they capture the linguistic reality of cultural and communicative memory. For an interdisciplinary research, they can shed light on the societies of the camps (Garbe 2015, p. 268 ff.) and/or contribute to a praxeology of language and memory in the sense of "doing memory" (Binnenkade 2015), also with a view to emotions and historical learning (Lange 2011; Heyl 2013). The empathetic attitude with which the interviewers encounter Nordheim, as well as the similarity of the utterances with which the guide passes down his knowledge and experiences, indicates, in my opinion, a high intrinsic motivation of these memory specialists: Social commemoration, the assumption and transmission of testimony as "secondary testimony" (Hartman 2000) require solidarity; they are ultimately forms of cultural and political action.

Appendix 1

Archive—Interview with Nordheim 2000, ((00:21:20–00:24:48)).

[1]

Int	[Da hab ich/ äh/]		Wie war das, wenn Sie dort stundenlag stehen mussten als Kind,
Int English	I/ äh		How was it when you had to stand there for hours as a child, that's
commentary	[deiktic "da"; possibly in the sense of "in this respect"		

[2]

MN		Schrecklich!	
MN English		Terrible!	
Int	das ist ja noch was anderes	als als Erwachsener? Als Erwachsener weiß man, man muss das
Int English	something else	than as an adult?	As an adult, you know you have to do it now, but children are often/

[3]

Int	jetzt tun, aber Kinder sind ja dann oft/ die wollen dann nicht · · oder die müssen. Was ist dann
Int English	they don't want to - - or they have to.	What happened then?

[4]

MN		(Passiert), · · wir haben das gewusst, dass wir keine andere (Chance) haben. Kinder
MN English		(Happened), · · we have known that we have no other (chance).	Children also find
Int	passiert?	
Int English		

[5]

MN	finden auch andere Sachen. Ich kann/ ich geb Ihnen/ ich geb Ihnen ein/ drei/ oder drei ähm, ähm · ·
MN English	other things to do.	I can/ I give you/ I give you one/ three/ or three ähm, ähm - - I can tell you three stories.

[6]

MN	kann ich Ihnen drei Geschichten erzählen. Eins: Wenn es sehr kalt war, wir spielen, ich und meine
MN English	One: When it was very cold, we play, me and my sister

[7]

MN	Schwester [((klatscht in die Hände))]˙ Kennen Sie das?	Die Handen warm zu machen. Und
MN English	[((claps hands))]	Do you know this?	Warming up the hands.	And then
Int		Hm˙	
commentary	[child's game, counting rhyme or similar		

[8]

MN	dann, wenn jemand kam, die SS oder (Albala) oder weiß ich nicht [((Hepp!))]	Oder in
MN English	when someone came, the SS or (Albala) or I don't know	[(Hepp!)]	Or in another
commentary		[gesture; "Attention"; adopts posture	

[9]

MN	einem anderen Haus, jemand sagte: "Sie kommen!". (Waren wir schnell weg). Kinder finden immer
MN English	house, someone said:	"They're coming!".	(We were gone quickly).	Children always find something

[10]

MN	was zu tun! Ich kann mich erinnern, ich war auf Appell · · · und ich musste · · äh mein/ wie sagt man/
MN English	to do!	I can remember, I was on roll call - - and I had to - - äh give my/ how do you say/ äh my urine?

[11]

MN	mein Urin äh geben? Ja, Pipi machen, ja. ((2s)) Sò, das war schrecklich. Sagt meine Mutter · · ·
MN English		Yes, make wee-wee, yes.	((2s)) So, that was terrible.	Says my mother · · · mother says:
Int		Pipi.	

Int English	Wee-wee

[12]

MN	Mutter sagt: "Mach in/ in · · deine Hose!". "[Ich kann nicht! Ich kann nicht, (ja)].". (Unerträglich). "[Ich
MN English	"[Just wee] - - your pants!". "[I can't! I can't. (yes)].". (Unbearable). "[I can't
commentary	[child voice, gesticulating] [child]

[13]

MN	kann nicht in die Hose machen, (das geht)/ kann ich nicht!. Ich muss, ich muss, ich muss]!". Und das
MN English	wet my pants, (I can)/ I can't. I have to, I have to, I have to]!". Und das war
commentary	voice, gesticulating Insistently

[14]

MN	war zwischen [(ein Zehn und anderes Zehn)], ich weiß nicht was · · · und dann all diese Frauen
MN English	zwischen And that was between [[one ten and other ten]] I don't know what · · · and then all these women
commentary	possibly means: between columns of ten prisoners

[15]

MN	haben [((1,5s))] um mich hin · · haben sie einen Mauer gemacht, so und ich hab P/ Pipi gemacht.
MN English	[(1,5s)] around me · · they formed a wall, so and I did P/ pee.
commentary	[gesture: circle]

[16]

MN	[Und das war, das war lang! Und meine Mutter war nervös, schrecklich! Das · hat kein, hat kein
MN English	[And that was, that was long] And my mother was nervous, terrible! That · has no, has no and and that goes
commentary	[smiles, undefines gesturally]

[17]

MN	Ende und das geht weiter und es war schrecklich! Ich kann mir das erinnern bis heute, bis heute!
MN English	on and it was terrible! I can remember that to this day, to this day!

[18]

MN	Das war schrecklich]! (Das sind diese/ man muss diese Sache verstehen, das ist, das/ kann man
MN English	That was terrible]! (That's this/ you have to understand this thing, that's, that/ you can't/ understand that/ what was there/

[19]

MN	nicht/ das/ verstehen, was da war! Und · · · äh · · · das Schreckliche war, das auch in/ in diese
MN English	And · · · äh · · · the terrible thing was that also in/ in those years, those clothes were terrible.

[20]

MN	Jahre, diese Kleider schrecklich waren. Schuhe waren nicht mehr Schuhe. Das war Karton. Das war
MN English	Shoes were no longer shoes. That was cardboard. That was

[21]

MN	schon · · äh · · schrecklich. Wir haben ein/ wir haben kalte Füße gehabt, wir haben nasse Füße
MN English	already · · äh · · terrible. We have had a/ we have had cold feet, we have had wet feet · · and the clothes were not clothes anymore.

[22]

MN	gehabt · · und die Kleider waren nicht mehr Kleider. (Auch) nicht kein neue Kleider. (Das) alles war ·
MN English	(Also) not no new clothes. (That) everything was ·

[23]

MN	schrecklich, war es. Es war ((unverständlich 1s)). So, war nicht warm, (es) war nicht zu glauben!
MN English	*terrible, it was.*	*it was ((unintelligible 1s)).*	*So, was not warm, (it) was not believeable!*

[24]

MN	So, wenn/ wenn man steht auf Appell, auf Strafappell • • • Stunden! Fünf, sechs, sieben, acht Uhr
MN English	*So, when/if one stands on roll call, on*	*punishment roll call • • • Hours!*	*Five, six, seven, eight o'clock • • • one stands*

[25]

MN	• • • steht man dort — Stunden —, steht man dort • • • und ist kalt. Viele Menschen sterben dann.
MN English	*there — hours —, one stands there • • • and is*	*cold. Many people die then.*

[26]

MN	ENDE ABSCHNITT 2 ((00:24:48:18))
MN English	*END OF SECTION ((00:24:48:18))*

Appendix 2

NDR—Interview with Nordheim 2015, ((00:26:49–00:30:42)).

[1]

Int	ANFANG ((00:26:49))	Herr Nordheim, ich möchte noch mal genauer wissen · · ähm:
Int English	BEGINNING ((00:26:49))	Mr. Nordheim, I would like to know again in more detail · · ähm:
Kommentar	MV spricht mit deutlichem holländischem Akzent	
commentary	MV speaks with a clear Dutch accent	

[2]

		fragend		
Int	Sie waren damals · neun	Jahre alt?	Oder zehn Jahre alt?
Int English	you were · nine	years old at that time?	Or ten years old?
MN		Ich war/ jä. Ich...	(Ich) war
MN English		I was/ yes. I...	(I) was in
commentary	interrogative emphasis		

[3]

Int		Zehn.	Hmhm·	Können Sie noch
Int English		Ten.	Hmhm·	Can you tell me again a
MN	in Bergen-Belsen, war ich schon zehn...	Zehn.	Zehn. Schon zehn.	
MN English	Bergen-Belsen, I was already ten...	Ten.	Ten. Already ten.	

[4]

Int	mal ein bisschen, wenn/ weil (wir) jetzt gesagt haben "Beispiele", zäh/ erzählen Sie mir bitte noch
Int English	little bit, if/ because (we) have now said "examples", te/ tell me again please · · a little bit more precisely, what did it mean to be in a · ·

[5]

	betont, lauter
Int	mal · · ein bisschen genauer, was hat das bedeutet, als zehnjähriges Kind ((1s)) in einem · ·
Int English	concentration camp as a ten-year-old child ((1s))?

[6]

Int	Konzentrationslager zu sein?		
MN		Ich bin erwachsen [(worden)]/ geworden. Ich bin kein Kind gewesen.
MN English	I	[(became)]/ grew up. I was not a child.
Kommentar		(holländisch)
commentary		(Dutch)

[7]

MN	Von die Zeiten, das die Krieg angefangen (bin), bin ich kein Kind mehr/ äh Kind mehr gewesen.
MN English	From the time the war started, I was no longer a child.

[8]

MN	Ich hab Situa/ Situationen sehr gut verstanden und ich hab gehandelt wie ein Erwachsener. Und ich
MN English	I understood situations very well and I acted like an adult. And I know

[9]

MN	weiß, ich hab kein/ ich hab nicht, · was, was die englische Menschen sagen [childhood] ha/habt, hab
MN English	I didn't have/ I didn't have, · what the English people say [childhood] ha/had, I never
Kommentar	(engl. Kindheit)

[10]

MN	ich nie gehabt! Hab ich nie gehabt! ((1,5s)) Ich weiß das! (Nur) wa/ wa/ was kann ich machen?
MN English	had! I never had! ((1,5s)) I know that! (Just) wh/ wh/ what can I do?

[11]

		langsamer
MN	((3s)) Und dann ((2s)) [hab ich zwei Sachen zu sagen]: In so ein [((unverständlich 1s))] In so ein	
MN English	((3s)) And then ((2s)) [I have two things to say]: in such a [((unintelligible 1s))] in such a situation	
Kommentar	lacht kurz	verhaspelt sich
commentary	laughs briefly	snarls

[12]

	stockt	
MN	Situation in/ äh/ in/ in/ in/ in/ in/ • in Appell: ((1,5s)) Es war sehr kalt ((1s)) (und meine Schwester und	
MN English	in/ äh/ in/ in/ in/ in/ in/ • in roll call: ((1,5s)) It was very cold ((1s)) (and my sister and I • • wanted to make our	

[13]

MN	mich • • wollten unsere Han/ Hande ein bisschen • • warmer machen). Haben wir gespielt • • und
MN English	hands/ hands a little • • warmer). Did we play • • and the woman was

[14]

		lauter	
MN	die Frau war sehr • nervös. [((1,5s))]	Oder: [((0,7s))] Ja? (Já, sie war nervös)!	
MN English	very • nervous. [((1,5s))]	Or: [((0,7s))] Yes? (Yes, she was nervous)!	
Kommentar	klatscht in die Hände wie beim Kinderspiel	reibt die Hände	
commentary	claps hands as if playing children's game	rubs hands	

[15]

	ruft
MN	(Aber waren wir Kinder)! Trotzdem dass ich erwachsen war, waren (wir) auch Kinder, wir wollte
MN English	(But we were children)! Even though I was an adult, (we) were also children, we wanted to have

[16]

		gedehnt
MN	[• • •] warme Hande habe! Das hab ich auch gemacht. Und dann hab ich eine andere	
MN English	[• • •] warm hands! I did that, too. And then I have another Episode:	
Kommentar	klatscht in die Hände	
commentary	claps hands	

[17]

MN	Episode: ((1,5s)) Ich musste • • • [Urine] geben. (Já), was kann ich tun? _Ich war ein Kind. [((2s))]
MN English	((1,5s)) I had to give • • • [Urine] (Yes), what can I do? _I was a child. [((2s))]
Kommentar	flüst. Urin	Mimik:
commentary	[Dutch: Urine]	[facial]

[18]

MN	Und ich sage sehr leicht zu meiner Mutter: "Ich kann nicht mehr! Ich kann nicht mehr
MN English	And I say very easily to my mother: "I can't hold anymore! I can't hold any more!".
Kommentar	machtlos
commentary	expressions: powerless

[19]

MN	halten!". "[Mach in de Hosen!]". [((1s))] Ich will sehr/ ich war sehr/ ich sag: "Nee!
MN English	"[Wet your pants!]". [((1s))] I want very much/ I was very much/ I say: "Nah! I/ I
Kommentar	[Mimik: ärgerlich, bedrängt] schüttelt entschieden den Kopf
commentary	[facial expressions: angry, distressed] shakes head resolutely

[20]

MN Ich/ Das mach ich nicht!". "[Aber wir können nicht anders!]". Ich hab die Mä/ die Frauen (da um/ um-
MN English don't do that". "[But we have no choice!]". I have said to the gi/ the women (there an/ aro- • •-und)
Kommentar [Mimik; Redewiedergabe der Mutter
commentary [facial expressions: mother's rendition of speech

[21]

MN • -her) • • hab ich ge/ gesagt: "Ich kann nicht mehr!". Und die Frauen haben einen Kreis gemacht um
MN English • • I have ge/ said: "I can't anymore!". And the women made a circle around me • • • and I gave urine.

[22]

MN mich (hin) • • • und ich hab Urine gegeben. Und das war lange! Das war lang! Das war lang! [Und sie
MN English And that was long! That was long! That was long! [And they
Kommentar [lacht
commentary [laughs

[23]

            lauter   ruft
MN sind nervös geworden, ich br/] • • versteh das heute! Damals hab ich nicht verstehen! Ich verstehe
MN English got nervous, I/ ...]   • • understand that today! I didn't understand then!   I understand today
Kommentar
commentary

[24]

MN heute, wie nervös sie waren. Ja, ich habe geguckt, das ni/ • • auch weit • • kein gefährliche Sache
MN English how nervous they were. Yes, I looked, that no/ • • also far • • was no dangerous thing, • • • but • • • it's a bit

[25]

            betont
MN war, • • • aber • • • es ist ein bisschen komisch, aber das [((unverständlich 1s))]/ Ich kann (mir)
MN English funny, but that [((unintelligible 1s))]/ I can see • how • • a
Kommentar [verhaspelt sich
commentary [mumbles

[26]

MN sehen, • wie • • ein schreckliche Welt das war! Ich konnte nicht (ein Urine) geben, ohne Strafe! ((2s))
MN English terrible world that was! I could not give (a urine) without punishment! ((2s))

[27]

MN Es ist glücklich, alles gut verlaufen, aber ich • verst/ ich erzäh| das [((1s))] um ein bisschen die
MN English Everything went well, but I - unders/ I tell that [((1s))] to give a little impression, wat
Kommentar Gestik: Hände
commentary gestures: hands

[28]

MN Eindruck zu machen, wat das/ Wat das heißt! ((2s)) Das ist • • Appell, (das ist eine schreckliche
MN English that/ What that means! ((2s)) That's • • roll call, (that's a terrible thing), (then) you also have to

[29]

MN Sache), (dann) muss man auch noch sagen: Kleider! Was für Kleider hab ich gehabt? (Mit den
MN English say: clothes! What kind of clothes did I have? (With the clothes

[30]

MN Kleidern, die wo komisch sind). [((schmunzelt))] Das waren keine Kleider mehr, das

MN English	that were funny).	[((smirks))]]	Those were not clothes anymore, that were/ was
Kommentar		[Mimik: Ohnmacht, Resignation	
commentary		[facial expressions: powerlessness, resignation	

[31]

			ruft
MN	seien/ war • das/ (Man kann das • nennen, äh • weiß ich was) • äh • war das • • • [Schmátes]!		
MN English - that/	(You can • call that, äh • I know what) • äh • was that • • •		[Schmátes]!
Kommentar			[österreich. Jiddisch: Lumpen,
commentary			[Austrian Yiddish: rags, scraps

[32]

MN	((1s)) Nichts! (D/ waren das neue Kleider, oder so)? [Es war auch • nicht mehr warm. Und es
MN English	((1s)) Nothing! (W/ were those new clothes, or something)? [It was also • no longer warm. And it was also no longer •
Kommentar Felien	[Gestik: zählt auf
commentary	[gestures: counts up

[33]

MN	war auch nicht mehr • • äh • • gegen Regen. • • • nix]! [Shoes] • • • meine • • Schuhe. • • das war
MN English • äh • • against rain, • • • nothing]!	[Shoes] • • • my • • shoes, • • had become
Kommentar	
commentary	[engl. Schuhe

[34]

	betont		
MN	Karton gew/ geworden. Das war nicht mehr kein Schuhe! Und das [muss man alles, muss man
MN English cardboard	That was no longer no shoes!	And that	[you have to take everything (in account)]
Kommentar			[Gestik: vermischt alles mit den Händen
commentary			[gestures: mixes everything with the hands

[35]

MN	beinehmen], dass man/ man lebt, • nicht mehr in gewöhne /-wohne Welt/ Welt. ENDE ((00:30:42))
MN English	that you/ you live, • no more in the accustomed world/ world. END ((00:30:42))
Kommentar	
commentary	

Appendix 3

Tour—Passing down Nordheim's testimony 2012, ((00:25:30–00:26:30)).

[1]

Guide	ABSCHNITT 4 ((00:25:35)) Sö˙ Wir sind jetzt hier. An dieser Stelle befand sich zur
Guide english	SECTION 4 ((00:25:35)) Sö˙ We are here now. At this place, in camp time,
Gruppe	
Group	//(unverständlich)//
Kommentar	//(uninteligible)//
commentary	[SKL mehrere; Nebendiskurse [School class several; secondary discourse

[2]

Guide	Lagerzeit, [das heißt ((1s)) als hier] Konzentrationslager war, • • war hier der Appellplatz
Guide english	[that is ((1s)) when there was] concentration camp here, • • here was the roll call place of the Star
Kommentar	[blickt ermahnend; beendet Nebendiskurse
commentary	[looks admonishingly; ends secondary discourse

[3]

Guide	des Sternlagers. Appelplatz, es war ein riesengroßer Platz und die Leute mussten morgens und
Guide english	Camp. . The roll call site was a huge square and the people had to line up in the morning and in the evening for the

[4]

Guide	abends zum Apperel/ Appell antreten, der in der Regel zwei bis drei Stunden dauerte. In der
Guide english	ro/ the roll call, which usually lasted two to three hours. In the so-called

[5]

Guide	sogenannten Habachtstellung, ohne Rücksicht darauf, ob es ein Kind ist oder ein Erwachsener.
Guide english	attentive positon, without regard to whether it was a child or an adult.

[6]

Guide	Es wurde gezählt, • • noch mal gezählt • • bis die Zahl stimmte — reine Schickanemaßnahme —
Guide english	It was counted, • • counted again • • until the number was right — mere chicanery —

[7]

Guide	und in dieser Situation ist der zehnjährige Moshe Nordheim, der mit seiner achtjährigen Schwester
Guide english	and in this situation is ten-year-old Moshe Nordheim, who is here with his eight-year-old sister and with his parents.

[8]

Guide	hier ist und mit seinen Eltern. Und er steht auf dem Appellplatz und die menschliche Situation:
Guide english	And he's standing on the roll call square and the human situation:

[9]

Guide	quengelig hart
Guide	"[Mama, ich muss mal.". Und er quengelte in einer Tour und die Mutter sagte dann: "Mach dir
Guide english	"[Mom, I have to pee.". And he kept whining and then the mother said: "Pee your
Kommentar	[stilisiert Dialog; deutlich betont
commentary	[restages dialogue; clearly emphasized

[10]

Guide	quengelig
Guide	in die Hose!". _"Aber ich kann mir nicht in die Hose machen!]". Und die Mutter war schon ganz
Guide english	pants!". _."But I can't pee my pants!]". And the mother was already quite panicked • •

[11]

Guide	panisch • • und hat die Situation genutzt, dass die Wachen am anderen Platzende sind und die

| Guide english | and took advantage of the situation that the guards are at the other end of the square and the woman around formed/ formed a circle |

[12]

| Guide | Frauen in der Umgebung bilten/ bildeten einen Kreis und der Moshe erzählt in seinem Interview: |
| Guide english | and Moshe tells in his interview: |

[13]

| Guide | "[Und ich stellte mich in die Mitte und es lief und lief und die Frauen wurden schon ganz ungeduldig. |
| Guide english | "[And I stood in the middle and it ran and ran and the women were already getting quite impatient. |

[14]

	lauter		
Guide	Und erleichtert stellte ich mich zurück und keiner hat's gesehen!]".	Wichtig.	Warum?
Guide english	And relieved, I stood back and no one had seen it]".	Important.	Why?
Schüler 2			((3,3s)) Da
S2 English			((3,3s)) Since that

[15]

Schüler 2	das • gang und gäbe war,	bei denen, • • schätz ich mal.	Also (die • kannten
S2 English	• was commonplace,	with them, • • I guess.	So (they • all knew that
Gruppe		[((1,1s)) Wed was]?	[((Warum, meint er]?
Group		[((1,1s)) Because what]?	[(Why, he thinks]?
Kommentar		[Schülerin A: unbekannt	[Schülerin B: unbekannt [Schülerin A
commentary		[Student A: unknown	[Student B: unknown [Student A

[16]

Guide	Okay.	
Guide english	Okay.	
Schüler 2	das • •	alle schon).
S2 English	• •	already).
Schüler 8		Äh halt die Wachen/ also, die mussten halt alle stehenbleiben und (wer)
S8 English		Äh well the guards/ they all had to stop and (who) didn't, they were (simply shot).
Gruppe		
Group		
Kommentar		
commentary		

[17]

Guide		Erschossen, ähm, oder eben bestraft, für die
Guide english		Shot, ähm, or just punished, for littering the place.
Schüler 8	nicht, der wurde (halt einfach erschossen).	
S8 English		

[18]

Guide	Verunreinigung des Platzes.	Es ist so, dass ((2,3s)) es • unterschiedliche Strafen gab. Die
Guide english		In fact ((2,3s)) there were • different punishments. The worst
Schüler 8		(Jä)˙
S8 English		(Yes)˙

[19]

| Guide | schlimmste ist Erschießen, ist aber nicht so oft vorgekommen, muss ich sagen, man hat andere |
| Guide english | is shooting, but it didn't happen so often, I must say, they found other wa/ äh ways: |

[20]

Guide	Wel/ äh Wege gefunden: Essensentzug, ** Stehen, * am Zaun stehen wenn man umkippt, dass
Guide english	Food deprivation, ** standing, * standing by the fence if you fall over, that you fall into the fence.

[21]

Guide	man in den Zaun fällt. Oder es gibt Wachleu/ äh Wachman/ -männer oder -frauen, die haben genau
Guide english	Or there are guar/ äh guards (men or women), they saw exactly,

[22]

	betont
Guide	gesehen, ** das sich die Leute in die Hose machen, ich mein, bei Durchfallerkrankungen, * das lief
Guide english	** that people wet their pants, I mean, with diarrhea, * it ran down the legs!

[23]

Guide	die Beine runter! Und dann gab's den Befehl für die gesamten Leute, die hier standen, Hinlegen und
Guide english	And then there was the order for all people who were standing here, to lie down and

[24]

	betont	gedehnt
Guide	Aufstehen, dass man sich entsprechend in dieser/ in diesem Dreck wälzt. Bei Regen, bei Hitze,	
Guide english	get up, so they would roll around in this dirt. in rain, in heat, always	

[25]

Guide	immer muss man hier stehen und: Im Winter ist es vorgekommen, dass Frauen in Sommerkleidern
Guide english	you have to stand here and: In the winter it happened that women in summer dresses stood here and men or boys

[26]

Guide	hier standen und Männer oder Jungs in kurzen Hosen, weil sie im Sommer verhaftet worden sind.
Guide english	in short pants, because they were arrested in the summer.

[27]

	eindringlich
Guide	Da hat man keine Rücksicht drauf genommen!
Guide english	There was no consideration for it!

References

Assmann, A. (2006). *Der lange Schatten der Vergangenheit*. München: Beck.

Assmann, J. (1992). *Das kulturelle Gedächtnis. Schrift, Erinnerung und politische Identität in frühen Hochkulturen*. München: Beck.

Ballis, A. (2019). Guides an KZ-Gedenkstätten und Holocaust Museen—Professionalisierung in Zeiten eines Wandels der Erinnerungskultur. In A. Ballis, & M. Gloe (eds.): *Holocaust Education Revisited. Wahrnehmung und Vermittlung—Fiktion und Fakten—Medialität und Digitalität* (141–166). Wiesbaden: Springer.

Binnenkade, A. (2015). Doing Memory. Teaching as a Discursive Node. *Journal of Educational Media, Memory and Society*, 7(2), 29–43.

Bock, D. (2017). *Literarische Störungen in Texten über die Shoah*. Frankfurt a. M.: Peter Lang.

Bredel, U. (1999). *Erzählen im Umbruch*. Tübingen: Stauffenburg.

Brünner, G. (1991). Redewiedergaben in Gesprächen. *Deutsche Sprache,* 1–15.

Ehlich, K. (1982). Deiktische und phorische Prozeduren beim literarischen Erzählen. In E. Lämmert (ed.), *Erzählforschung* (112–129). Stuttgart: Metzler.

Ehlich, K. (1983). Text und sprachliches Handeln. Die Entstehung von Texten aus dem Bedürfnis nach Überlieferung. In A. Assmann et al. (eds.), *Schrift und Gedächtnis. Beiträge zur Archäologie der literarischen Kommunikation* (24–43). München: Fink.

Ehlich, K. (1997). Medium Sprache. In H. Strohner, L. Sichelschmidt, & M. Hielscher (eds.), *Medium Sprache* (9–21). Frankfurt a. M.: Peter Lang.

Ehlich, K., & Rehbein, J. (1976). Halbinterpretative Arbeitstranskriptionen (HIAT). *Linguistische Berichte* 45, 21–41.

Ehlich, K., & Rehbein, J. (1977). Wissen, kommunikatives Handeln und die Schule. In: H.C. Goeppert (ed.), *Sprachverhalten im Unterricht. Zur Kommunikation von Lehrer und Schüler in der Unterrichtssituation* (36–115). München: Wilhelm Fink.

Ehlich, K., & Rehbein, J. (1986). *Muster und Institution. Untersuchungen zur schulischen Kommunikation.* Tübingen: Narr.

Garbe, D. (2015). *Neuengamme im System der Konzentrationslager. Studien zur Ereignis- und Rezeptionsgeschichte.* Berlin: Metropol.

Gryglewski, E., Haug, V., Kößler, G. et al. (eds.) (2015). *Gedenkstättenpädagogik. Kontexte, Theorie und Praxis der Bildungsarbeit zu NS-Verbrechen.* Berlin: Metropol.

Gudehus, C. (2006). *Dem Gedächtnis zuhören. Erzählungen über NS-Verbrechen und ihre Repräsentationen in deutschen Gedenkstätten.* Essen: Klartext.

Hartman, G. (2000). Intellektuelle Zeugenschaft. In U. Baer (ed.), *Niemand zeugt für den Zeugen. Erinnerungskultur nach der Shoah* (35–53). Frankfurt a. M.: Suhrkamp.

Haug, V. (2015). *Am "authentischen" Ort. Paradoxien der Gedenkstättenpädagogik.* Berlin: Metropol.

Heyl, M. (2013). Mit Überwältigendem überwältigen? Emotionen in KZ-Gedenkstätten. In J. Brauer, & M. Lücke (eds.), *Emotionen, Geschichte und historisches Lernen. Geschichtsdidaktische und geschichtskulturelle Perspektiven* (239–261). Göttingen: Vandenhoeck & Ruprecht Unipress.

Jürgensen, F. (2003). Unter dem Teppich liegt der Standard. Wie kommt es, daß die Führung in Gedenkstätten so wenig bearbeitet wird? *Standbein Spielbein,* 65, 55–60.

Kabalek, K. (2015). Edges of History and Memory. The Final Stage of the Holocaust. *Dapim. Studies on the Holocaust,* 29(3), 240–263.

Lakoff, G., & Johnson, M. (1980). *Leben in Metaphern. Konstruktion und Gebrauch von Sprachbildern.* Heidelberg: Carl Auer.

Lange, K. (2011). *Historisches Bildverstehen oder Wie lernen Schüler mit Bildquellen?* Berlin: LIT.

Lautenbach-von Ostrowski, M. (2014). "Heut ist das 'ne Jugendherberge." Eine Fallstudie zu sprachlichen Wissensvermittlungsformen bei einer Schülerführung in der KZ-Gedenkstätte Ravensbrück. In N. Benitt, C. Koch, K. Müller et al. (eds.), *Kommunikation—Korpus—Kultur. Ansätze und Konzepte einer kulturwissenschaftlichen Linguistik* (109–141). Trier: wvt.

Lautenbach-von Ostrowski, M. (2015). "Ich bin heute hier und ihr nicht mehr!" Linguistische Fallanalyse von Zweck und Funktion einer Redewiedergabe in einer KZ-Gedenkstätten-Führung. In M. Becker, D. Bock, & H. Illig (eds.), *Orte und Akteure im System der NS-Zwangslager* (256–278). Berlin: Metropol.

Lautenbach-von Ostrowski, M. (2020). "Man behandelt sie bewusst wie Tiere." Sprachliche Zugänge zur Erinnerung bei Führungen in KZ-Gedenkstätten. In M. Gloe, & A. Ballis (eds.), *Holocaust Education Revisited. Orte der Vermittlung—Didaktik und Nachhaltigkeit* (15–37). Wiesbaden: Springer.

Lautenbach-von Ostrowski, M. (forthcoming, a). *Sprachliches Handeln als Zugang zur Erinnerung. Schülerführungen in KZ-Gedenkstätten.* Doctoral thesis, Faculty of Humanities at the University of Hamburg (2018).

Lautenbach-von Ostrowski, M. (forthcoming, b). Going Beyond Narration. Guided Tours at Concentration Camp Memorial Sites from a Linguistic Perspective. *Narrative Culture.*

Levi, P. (1986). *Die Untergegangenen und die Geretteten.* München: Hanser.

Pampel, B. (2007). *Mit eigenen Augen sehen, wozu der Mensch fähig ist.* Frankfurt a.M.: Campus.

Redder, A. (2003). Literarische Kommunikation. In J. Hageman, S. Sager, & F. Svend (eds.), *Schriftliche und mündliche Kommunikation: Begriffe—Methoden—Analysen* (185–197). Tübingen: Stauffenburg.

Redder, A. (2008). Functional Pragmatics. In G. Antos, & E. Ventola (eds.), *Handbook of Interpersonal Communication* (133–178). Berlin: Mouton de Gruyter.

Redder, A., & Rehbein, J. (1987). Zum Begriff der Kultur. *Osnabrücker Beiträge zur Sprachtheorie. OBST*, 38, 7–22.

Rehbein, J. (1977). *Komplexes Handeln. Elemente zur Handlungstheorie der Sprache.* Stuttgart: Metzler.

Rehbein, J., Schmidt, T., Meyer, B. et al. (2004). *Handbuch für das computergestützte Transkribieren nach HIAT. Arbeiten zur Mehrsprachigkeit—Folge B*, Nr. 56/2004. Universität Hamburg: SFB 538 Mehrsprachigkeit.

Zülsdorf-Kersting, M. (2007). *Sechzig Jahre danach: Jugendliche und Holocaust. Eine Studie zur geschichtskulturellen Sozialisation.* Berlin: LIT.

Informing, Accompanying, Commemorating at the Memorial Site Dachau

Teacher Guides' Experiences and Reflections

Michael Penzold

Abstract

At the Memorial Site Dachau teachers can work as guides for a limited time supported by the Ministry of Education in Bavaria. How teachers experience this change from the classroom to the memorial is analyzed with the help of qualitative interviews and ethnographic observations. Three questions are of central concern, formulated from the perspective of the teacher guides: How do I change as a teacher when I become a guide at the memorial site? What am I like as a teacher when I have had more experience working as a guide? What have I learned about myself based on my memorial work routines? To answer these questions, the author describes three stages of professional development which are grounded in data. These stages offer insights into shaping the the educational work on site.

1   Teacher Guides at Memorial Sites

Students and teachers in Germany often study the Holocaust multiple times and in different school subjects. They often come to realize that their relatives who lived at the time of National Socialism may have participated in activities carried out by this regime. They therefore ask questions about this time and use different

M. Penzold (✉)
Institut für Germanistik, Ludwig-Maximilians-Universität München, Munich, Germany
e-mail: Michael.Penzold@germanistik.uni-muenchen.de

© The Author(s), under exclusive license to Springer Fachmedien Wiesbaden   71
Gmbh, part of Springer Nature 2022
A. Ballis (ed.), *Tour Guides at Memorial Sites and Holocaust Museums*,
Holocaust Education – Historisches Lernen – Menschenrechtsbildung,
https://doi.org/10.1007/978-3-658-35818-1_5

media to find answers to these questions. It is not surprising that visits to the memorial site, which are very frequent today, have existed since the early days of the concentration camp memorial site.

For many visitors, concentration camps are the epitome of German tyranny in the age of National Socialism. Usually, students in Bavaria and from other countries visit both the grounds of the former Dachau concentration camp and the exhibitions. Since the 1980s, "school classes even make up the majority of visitors" (Hammermann and Pilzweger-Steiner 2017, p. 21). In order to cope with the increasing number of visitors, additional guides were recruited.[1]

To help fill the need for more guides, the Bavarian Ministry of Education has, in cooperation with the institution State Center of Political Education [Landeszentrale für politische Bildung] developed a program which allows teachers in Bavarian Schools to apply for the position as a teacher guide to guide school classes at the memorial site. While the practice of teachers guiding students at the Memorial Site Dachau is quite common, teacher guides are rare in the field of international memorial work in general. These teachers do not necessarily have to be history teachers. Teacher guides in Dachau are not obliged to participate in the training offered by the Memorial Site Dachau—in contrast to the guides who work as freelancers or volunteers. Ballis recently compared and discussed these training courses for guides at various institutions. Ballis emphasizes the importance of the training courses for guides in the context of standardization and professionalization (Ballis 2019; Ballis in this volume).

These factors do not seem to be important for teacher guides. Still unexperienced teacher guides often accompany more experienced teacher guides on their tours that have already been worked out. In this respect, in the context of this essay, they are to be called apprentices. It seems to be assumed here that learning to guide can be figured out without difficulty by experienced teachers. Further, it can be assumed that teachers who are interested in such an activity have a special personal or professional motivation. Teacher guides in this sense are teachers who also continue to teach in a school. As a result, physical working conditions change for them for one day a week. They exchange, at least temporarily, dry and heated classrooms for activities in fresh air, and they may even have to deal with a variety of conditions: In pouring rain, for example, they have to lead a group huddled under rain jackets and umbrellas (P44, 1, 30 f.). In this case, it can be difficult to hear one another as voices have to overcome the noise of the rain on the umbrellas and the crunch of the gravel under the students' feet (P44,

---

[1] The framework for this program is accessible on www.gedenkstaettenpaedagogik-bayern.de/hinweise.php.

2, 49 f.). In their new jobs, these teachers are now working in the public eye far more than before.

Teacher guides are visible to other teachers accompanying their school classes. During their visit at the memorial site, classes therefore seem to have two, perhaps even competing teachers: the teacher guide and their "own" teacher from the school accompanying the class. Occasionally these accompanying teachers may feel the desire to "take the lead" (L72, 7, 49–51). On the one hand, teacher guides cannot always expect that the accompanying teachers will support them, especially when the tour does not proceed as planned. If students are inattentive, distracted, or disruptive, the accompanying teachers may not always feel responsible (L84, 15, 29–31). On the other hand, there are teachers who are exclusively concerned with disciplinary rules for students who misbehave, instead of engaging in professional discourse or at least attentive listening (P47, 4, 13 f.).

Through their activities at the memorial site, teachers represent the official, ceremonial commemorative culture which is an integral part of Germany's national heritage. Together with visitors from all over the world, they participate in the internationally acclaimed wreath-laying ceremonies on the anniversary of the liberation on site (P44, 7, 37). In this way, they become part of a public, political form of event that is characterized to a large extent by security precautions—a factor that can lead to an additional burden on teachers in terms of time and organization (P44, 7, 37–44). The public interest goes hand in hand with distributing public and iconic symbols, in which, for example, it is already negotiable who lays which wreath where and when, and which pupil sits next to which eyewitness so that they may enter into conversation with one another (P44, 8, 42–51). Teachers who voluntarily work as guides are more exposed to public attention than they are in the classroom or school.

Of further interest are the teacher guides' attitudes towards their new field of activity. Teachers who guide tours at the Memorial Site Dachau have to guide school classes from all over Bavaria, and they have to take care of the students and their teachers visiting the site. Firstly, the teacher guides take on an active role by guiding school classes around the site. Secondly, they must also react sensitively to conditions on site. They have to change their program spontaneously if there is a crowd at certain points. They have considered the weather situation, too. They must be able to adjust to questions and moods of the school class they interact with. Their experience as teachers, acquired in the course of years of school teaching, allow them to do this with ease.

Based on my research on site, I found that these teacher guides may grapple with three basic questions:

- How do I change as a teacher when I become a guide at the memorial site?
- What am I like as a teacher when I get used to working as a guide?
- What have I learned about myself based on my memorial work routines?

It is interesting, though not surprising, that in answering these three questions, the teacher guides reflect on the memorial space and its consequences for guiding. Based on my data, I describe important aspects that emerge from the cognitive and practical dissonance between teachers and their new activity as guides. I can describe three stages of professional development. The stages are not fixed, but are connected with each other and related to the educational work on site.

Following grounded theory, the qualitative data (conversations, interviews, and field studies) were collected at the Memorial Site Dachau between 2015 and 2018. A total of 10 people contributed to the interviews.

2   Three Stages of Becoming a Teacher Guide

Stage I—Teachers change their role and experience themselves as apprentices at the memorial site
Teachers who apply for guiding usually do not live too far away from the memorial site; thus, are familiar with the location of the memorial. They have visited the site, and they know the nature of the guides' work from a visitor's perspective. Now as teacher guides they are students again and are also on the threshold of something new in a new space. They are on the way to a deeper understanding of a specific historical epoch and its consequences—which is why the term apprentice is appropriate at Stage I. One teacher guide explains:

> "Before I conducted the tour myself, I shadowed nine tours, some given by members of the State Center of Political Education in Bavaria, some by the Forum, and some by the memorial site—led by different people. It was always similar. And then you think, I'm ready to tell this story. There are also materials from the regional office that cover the essential points. Every colleague can actually download them, and every colleague could lead a school class independently across the grounds. The interesting thing is that when you're the tour guide, you read more deeply. And then somehow moments come along, that is, a deeper kind of insight" (L82, 7, 18–32; translated from the German).[2]

---

[2] "Bevor ich den Rundgang selber gemacht hab, glaub' ich, [hab' ich] neun Mal Rundgänge gehört, teilweise von der Landeszentrale für politische Bildung, teilweise vom Forum, teilweise von der Gedenkstätte von unterschiedlichen Leuten; es war immer immer immer ähnlich und gleich und dann denkt man so, ja, glaub' ich, trau' ich mir auch zu das zu erzählen

One interviewee begins by relating experience from many tours which he accompanied as a teacher. In addition, this guide emphasizes that at a certain point he realized that the various tours he participated in could well have been conducted by himself in a similar manner. This was the starting point for his application as a guide.

All the interviewees mention interest, knowledge, and a good opportunity as reasons for their decision to try something new.

The new teacher guide is now both student and initiate: The teacher mentioned above reports in an interview about the changes in everyday life. The experience of no longer being an ordinary teacher at school one day a week, but acting in a place of public interest, is important to him. He finds it exciting to work at a site of Nazi crimes, of all places, as a "Führer,"[3] a term with which the English word guide can be translated into German.[4]

"So I'm here on Thursday and I'm out then, I mean, considering all the practical issues, there are two guided tours, two tours. The term 'Führung' (guided tour) is loaded, you know. I am a tour guide in Dachau, so tour leader yes, o.k. so. I do these tours" (L82, 1, 31 f.; translated from the German).[5]

---

gibt ja auch Materialien von der Landeszentrale, zu was sind die wesentlichen Punkte, die kann sich jeder Kollege eigentlich runterladen und jeder Kollege könnte eigentlich ne Schulklasse eigenständig drüberführen. Das Interessante ist dann schon, dass man, wenn man jetzt der Rundgangsleiter ist, liest man doch vertieft und dann kommen doch irgendwie so Momente so Hinweise auf was Spezifisches also" (L82, 7, 18–32).

The speaker alludes here to tours offered by the Dachauer Forum Association (www.dachauer-forum.de/themen/gesellschaft-geschichte/guided-tours-dachau-concentration-camp-memorial-site/). Representatives of the State Center for Political Education in Bavaria [Landeszentrale für politische Bildungsarbeit] also offer specific tours in the context of state political and historical education projects and are the contact for the organization and recognition of class trips (www.blz.bayern.de/ueber-uns/verordnung.html).

[3] For Heine (2019), interestingly, the word "Führer" is not a word "burned" by National Socialist usage. This is surprising, since it is repeatedly cited as a prominent Nazi word in the context of the conversations with the teacher-guides. It is possible that in everyday usage rather certain composites of the word "Führer" are associated with National Socialism, such as "Führerbunker" or "Führerprinzip" (Simms 2019, p. 1031).

[4] The synonyms "Rundgangsleiter" or "Rundgangsleiterin" ("tour guide"), "Referentin" (f) or "Referent" (m) (lecturer) used at the memorial site are more appropriate to him (L82, 2, 2–4).

[5] "Also ich bin Donnerstag, und bin dann raus, ich mein', wenn man jetzt rein von den ganzen pragmatischen Geschichten her geht. Des sind zwei Führu- zwei Rundgänge der Begriff Führung ist ja belastet, kennt man ja. Ich bin ein Führer in Dachau, also Rundgangsleiter, ja, o.k., also ich mach' diese Rundgänge" (L82, 1, 31 f.).

The place where teachers work is characterized by an awareness of language use (Haug 2015). The new activity is quite different from the activity in the school, yet it is not unconnected to the way the Holocaust is talked about in schools. The interviewed teacher himself even emphasizes that he regularly and intensively dealt with the topic of the Holocaust in class (L82, 5, 13–17). However, just having begun working as a teacher guide (L82, 5, 7 f.), he distances his activity from his teaching at school. He emphasizes that he now is "out." This expression indicates that he understands "school" as a professional network of relationships, which is suspended for one day a week. The teacher guide becomes a liminal figure that exchanges a habitual, coveted, and hard-won membership in the teaching profession for a more fragile professional existence. Precariousness and liminality is also evident in self-designation, which reveals the spatial shift in language. While in everyday language it seems acceptable to refer to, for example, a guided city tour as a "Führung," it is not an appropriate expression for people working on site. The interviewed guides even apologized for mentioning the word "Führung" in their reports on memorial site visits (L63, 5, 33).

Furthermore, it is interesting for the teacher guide that by crossing the threshold, he seems to have entered another use and modulation of spoken language. This includes that "shouting is not really appropriate to the place" (L82, 2, 10 f.) at the memorial site, even though this might be useful for the accompanying teachers in the case of an undisciplined group for disciplinary reasons. A symbol of membership to the place is the "identity badge" (L82, 3, 33), which is worn so that it is clearly visible by all guides on site. This visual symbol is important for them since it legitimizes their position as a teacher guide. Nevertheless, they regard themselves as learners still, and not all places at the site are equally accessible for them. One teacher states:

> "Then it can happen that sometimes you just don't get this room. Some colleagues have connections to the memorial. [They] have [a] key, so this coordination between the National Center for Political Education and the memorial site management is a bit like that. You can tell who has been with us for a longer time and has somehow managed to get a hold of such a key—there is an adjoining room, an adjoining room seminar" (L82, 3, 17–26; translated from the German).[6]

---

[6] "Dann kann's passieren, dass man manchmal eben nicht diesen Raum kriegt. Einige Kollegen haben Beziehungen zur Gedenkstätte, haben Schlüssel, aber so diese Koordination zwischen Landeszentrale für politische Bildung und der Gedenkstättenleitung ist so ein bisschen, also da merkt man schon, wer schon irgendwie schon länger dabei ist und sich so einen Schlüssel ergaunert hat—es gibt so einen Nebenraum so einen Seminarnebenraum" (L82, 3, 17–26).

In this quote, the guide notes that he is new to the memorial and attempts to work out the logic of the space and the exhibition. He has not yet figured out how to get into the rooms that are closed to the public, which he regards as a disadvantage. This deficit could indicate that he perceives himself not as a fully accepted guide. Consequently, he tries to replace this deficient self-perception with historical content. He stresses the biographical component that shapes his new self-awareness of being a historian:

> "As a teacher, I always find it presumptuous to say that I am a historian, but now I have this one day when I am not at school. I can act like a historian, which is quite nice now and again. Sometimes, it's also really exciting" (L82, 12, 2–6; translated from the German).[7]

The quote illustrates that the teacher guide has individualized his role. Although he sees that he cannot enter all rooms equally, he can become a historian in the archive of the memorial. The understanding of being a historian is fundamental to his new role as teacher guide. In the following sequence of the interview, the metaphor of space is striking:

> "In guiding, I have the motivation and the luxury to deal with a topic in depth. So I've dealt with the Holocaust in the ninth grade all these years as a history teacher, eleventh grade too. It's of course obvious that you have to make it a topic. It's part of the curriculum, but the interesting thing is that I have to read more deeply into it. Well, now I notice that I have read a lot. I have come across many things again and again in the course of my private reading, but now I really have the feeling that the knowledge is contextualized" (L82, 5, 10–22; translated from the German).[8]

The reference to the memorial as a multi-layered and differentiated system of spaces and accesses is also important in the following remark, which then goes even further:

---

[7] "Als Lehrer ich find's immer vermessen zu sagen, ich bin Historiker, aber jetzt hab' ich halt diesen einen Tag, wo ich nicht in der Schule bin, kann ich mich mal so ein bisschen historisch gerieren. Und als Historiker ist es vielleicht auch ganz schön, es ist auch wirklich spannend" (L82, 12, 2–6).

[8] "Durch diese Abordnung so den Impetus hab' und den mir den Luxus gönne, mal wieder mich vertieft mit einer Thematik zu beschäftigen, also ich hab' all die Jahre als Geschichtslehrer natürlich in der neunten Klasse Holocaust behandelt, elfte Klasse logischerweise auch, das ist natürlich einleuchtend, dass man muss es thematisieren. Es ist Teil es ist Teil des Lehrplans, aber das, wie soll ich sagen, das Interessante ist jetzt schon, dass mich quasi vertiefter einlesen muss. Also ich merk', jetzt lesen gelesen hab'. Ich hab' Vieles immer wieder mal, ist man im Laufe seiner privaten Lektüre darüber gestoßen, aber so jetzt wirklich dieses eine Art so die an einem Ort verortet ist" (L82, 5, 10–22).

"And now you [teach] with the help of the place of learning, and when you are at the place of learning, then sometimes site-specific questions are asked, which simply do not arise in class. This is a completely different thing as every museum educator will tell you: There's a difference whether you are at the historical location or whether you are now bringing the historical location into the classroom by means of media or what have you. These are simply two different experiences" (L82, 6, 17–24; translated from the German).[9]

It is important for this teacher guide that the memorial is not just a possible classroom among many others. In his opinion, the spatial presence of the memorial site has a direct effect on the perception of the topic Holocaust. Because of the guides' physical presence at the memorial site and because of their experiences on site, the school classes can ask new, important questions or articulate emotions that even the best classroom cannot. He thus recognizes the category of spatiality as highly relevant didactically.

The teacher guide experiences his new activity which enriches his teaching in classrooms. He takes full advantage of the opportunity to work as a guide. Nevertheless, the apprentice is not simply being positive, but rather shows awareness of the complexity of tour guiding. He is also aware that teaching on site about the Holocaust and Nazi crimes requires a lot of psychological insight. For this reason, the guides should be free to work at the memorial only for a limited period of time: "You don't commit yourself to Dachau for a lifetime. I must also be able to say I cannot understand. This psychological handling of it, what the place is, is also special" (L82, 14, 30–33; translated from the German).[10]

At this point, one can summarize the characteristics of the Stage I of teacher guides' development: Starting from an activity as a teacher at a school, the new teacher guides begin to develop an awareness of the nuances of their new activity. This is important because they develop their tour narrative on their own by observing the tours of experienced colleagues. However, the space of the memorial seems to be a space segregated from the outside world by behavioral and linguistic taboos. The apprentice then decides to approach and expose himself to this space. A central part of this process is the realization that, as a teacher guide, he is making something

---

[9] "Und jetzt macht man's anhand eines Lernortes und das ist schon, wenn man am Lernort ist, dann werden da teilweise auch ortsspezifische Fragen gestellt, die sich im Unterricht einfach nicht ergeben. Das ist eine ganz andere [Situation], jeder Museumspädagoge wird ihnen sagen: Es ist ein Unterschied, ob man an dem historischen Ort ist oder ob man jetzt meinetwegen den historischen Ort medial ins Klassenzimmer bringt, das sind zwei unterschiedliche Momente" (L82, 6, 17–24).
[10] "Man verpflichtet sich nicht ein Leben lang nach Dachau, also es hat auch wirklich, es war wirklich, ich muss auch sagen, ich kann nicht verstehen diesen psychischen Umgang, damit also, was ist der Ort, ist schon auch speziell" (L82, 14,30–33).

possible that he could not accomplish in the same way in school. This includes the experience that the space of the memorial is multi-layered. In a very individual way, in the case studied, the apprentice seems to sense the possibilities and also the threatening nature of the space of the former concentration camp.

Stage II—Teacher guides develop language and behavior routines when accompanying other teachers and their classes
If teachers have dared to take the first steps into the memorial and start their new activity, further questions arise. While the apprentices of Stage I are still wondering about possible activities on site, the teachers in Stage II are interested in to what extent they can optimize their own educational program. It is important for the teacher guide at this stage to be able to "hear another voice" (L82, 15, 16), that is, to learn from and to be stimulated by other, more experienced memorial workers. Further, it is important to get advice from older and more experienced colleagues (L83, 1, 23 f.). But now it is more about working out a professional and pragmatic attitude as a guide. This can also include taking advantage of training offered by the memorial or other institutions.

After the stage of becoming familiar with the space of the memorial—in contrast to the situation in class—and after the first conducted tours, the guides consider practical and organizational issues. Colleagues encounter the new teacher-guides not only in their function as masters and instructors, but also at their level. After a number of observations and explorations, the teacher guides develop a routine practice, which they repeatedly improve and reflect.

Moreover, the number of successfully conducted tours also increases an awareness about how the guided school classes differ from each other. This applies not only to different German school types, but also to the emotional desire of the students to engage with the topic of the tour. The students, most of them aged between 14 and 17, come from sometimes very different types of schools. The differentiated German school system has an impact on the understanding of guiding. In addition, teachers in Germany are mostly trained for one type of school. For teacher guides, it is relevant to broaden their perspective and learn how to work with a group of heterogeneous students. With increasing routine, the teacher guide can react more flexibly to the personal requirements of the students.

However, the encounter with different students can also be inspiring for the teacher guides. They notice the quiet and probably respectful behaviour of a typical class at a German Gymnasium, a type of school which qualifies students to study at universities. One teacher guide stresses his excitement when constructing new practices. As a guide, he is surprised at the students' silence in contrast to their behaviour experienced during his school career:

"They weren't silent because of the subject matter, rather, I partly have the feeling that Gymnasium students are sometimes aware of the subject matter and of how they appear to others. There are some classes to which I always say: Hey guys, there are no grades here, I'm your expert. There are no stupid questions here. If you have questions, then ask them, that's what I'm here for. Sometimes it's just a matter of their self-image. Especially when it comes to this topic, they don't want to appear ignorant. Secondary school students, on the other hand, are surprisingly direct. They ask you some technical questions or questions about the heating in the bunker" (L82, 11, 2–15; translated from the German).[11]

The quote underlines that the guide already developed a professional self-image as a guide on the site and that he has already gained an overview of groups. This teacher understands himself as an expert and invites the students to ask questions. Further, his self-perception as an expert based on his experiences in guiding increases, and he wants to be very well prepared. He intends to avoid "weakness," for example, especially regarding students' questions: Can he, as a guide, correctly explain political decisions that led to the Holocaust? Can he convincingly explain to the students that it is important to deal with the Holocaust? Does he, as a guide, remember correctly the names of the historical persons he is talking about? He is also afraid to fail with "technical" questions about the camp: How many prisoners were in the camp, when, and for how long? How many SS men were used as guards?

Of course, no guide is required to know all the details. Nevertheless, it is obvious for the teacher guide that particular conscientiousness is required, especially when it comes to the Holocaust.

After a while, the teachers becoming guides get used to the local peculiarities of the memorial. Now they confidently decide which place to go first with their group and what to show or report there. The reflection on one's own route preference and those of other groups and guides, as well as the perception of the representational character of the memorial, is further developed in routine professional practice. The guides perceive themselves as a "help" (L82, 15, 11) for the teachers who only accompany the class. They introduce the memorial to the students and their teachers. For students who are listening, another space is created. It helps them to "see" what the place looked like in history. In Stage II, all interviewees consider the Holocaust as

---

[11] "Es war jetzt nicht die Stille irgendwie wegen der Thematik, sondern teilweise hab' ich das Gefühl, Gymnasiasten sind sich manchmal auch der Thematik und auch ihrer Außenwirkung bewusst. Also es gibt manche Klassen, da sag' ich immer, Hej Leute, hier gibt es keine Noten, ich bin euer Experte, ihr könnt hier gibt's keine blöden Fragen draußen, wenn ihr Fragen habt, stellt sie, dafür bin ich da. Aber da war's so ein bisschen, da war so dieses Selbstbild, was man nach außen hat, gerade bei der Thematik will man sich vielleicht keine Blöße geben. Hingegen Mittelschüler sind erstaunlich direkt, stellen einem keine Ahnung irgendwelche technischen Fragen oder irgendwie zur Heizung im Bunker" (L82, 11, 2–15).

a "safe" and uncontroversial topic of education. Their tour guiding follows routines that have been successfully carried out. Contradictions and conflicts are not in the guides' mind.

Nevertheless, the teacher guides are aware that teaching the Holocaust is something out of the ordinary and can lead to undesirable behavior of students at the memorial site. For the guides, such disturbances may be desirable because they interpret them as active involvement by the students. At school, they regarded such behavior as unacceptable, especially in their own lessons. On site, they can evaluate students' behavior as adequate reactions to a topic. Since it might be emotionally demanding for the students on site, the teacher guides are more tolerant: "Contrary to my own teaching, I don't hold it against any student if they then start to disengage with their neighbour when it comes to the subject matter or any pictures" (L82, 15, 31–34; translated from the German).[12]

As already mentioned, the differentiation of the visitor groups and their specific behaviour on site play a significant role in Stage II. Further, the teacher guides not only deal with school classes, but also with other groups of visitors they observe at the memorial:

> "I believe that it makes a difference whether I visit such a place as a survivor or as a relative of a survivor, or whether I am conveying something to young people having to do with political or historical education. That is one way of looking at it, but I think it's important to remember why the relatives are here. That's why I tell the students that when you go through the gas chambers, it should be clear to you where you are. People died there" (L 83, 5, 21–31; translated from the German).[13]

This statement illustrates the widening of the teacher guides' perspective. This interviewee regards himself as an agent of historical-political education. Since he is confronted with visitors who mourn at the memorial site, the teacher guide realizes that the memorial is not only a learning place, but also a cemetery. The relatives of former prisoners might see the place differently and connect it with relatives' lives. From the perspective of the teacher guides, they have to raise awareness and

---

[12] "Entgegen dem eigenen Unterricht nehme ich es keinem Schüler krumm, wenn er bei der Thematik oder bei irgendwelchen Bildern dann anfängt, irgendwie sich mit dem Nachbarn sich auszuklinken" (L82, 15, 31–34).

[13] "Ich glaub' das ist Unterschied, ob ich jetzt als Überlebender meinetwegen zum Beispiel noch mal so einen Ort besuche oder als Angehöriger eines Überlebenden oder ob ich das mache, als führe einen Jugendlichen im Sinne der politischen Bildung oder historischen Bildung über das Gelände. Das ist ein anderer Zugang, da muss der Zugang der Angehörigen [berücksichtigen] und deswegen ist es sinnvoll, das sag' ich den Schülern auch, das ist einleuchtend, wenn ihr da durchgeht durch die Gaskammern, bitte, einfach seid klar, wo ihr seid, da sind Menschen gestorben" (L 83, 5, 21–31).

empathy for the victims and their relatives. Consequently, the teacher guides shape their tours in a political and historical way, being mindful not to pass over the torture and pain people suffered on site.

In Stage II, the teacher guides also deal with the consequences of developing routines, for example, in observing other groups. The personal practice of guiding is embedded in the context of other tours. This might cause some conflicts, for example, with teachers of the classes, as the interviewee states:

"Sometimes, some of my colleagues [accompanying teachers] plan to conclude by watching the film in the theater at eleven-thirty. I've messed up a few colleagues' plans because I talked too long or the students asked too many questions. So far, I haven't heard any bad comments from any of my colleagues, but I haven't said anything more about it to them either. Maybe that's why some are miffed" (L83, 7, 3–12; translated from the German).[14]

In this quote, the teacher guide presents himself as a person who is particularly committed to his activity of talking to students. As the teacher above states, he is now confident that he will change the program if the conversations with the students go particularly well, even if he is in danger of exceeding the schedule of the tour.

In Stage II, the tour through the memorial appears to be a scarce commodity (L83, 9, 12 f.). The challenge for the guides is to use time in a perfect way. Although "access to the archive" (L83, 15, 29 f.) remains important in this phase, the guide's individual development as a historian seems to give way to a self-perception as a "performer." A quote from another interviewed teacher sums up Stage II: "If you wanted to emphasize a point somewhere, the other locations would be missing. It's context after all" (L83, 21, 32–34; translated from the German).[15]

While guides find their individual access to the topic and the memorial site in Stage I, they realize in Stage II that the memorial is a network of references and information. The teacher guides have to become part of this network and establish a routine for their tour guiding of visitors.

---

[14] "Es ist manchmal so, dass manche Kollegen auch extra letztendlich planen, dass sie anschließend noch um 11.30 Uhr sich den Film im Kino anschauen können. Ich hab's bisher einigen Kollegen vermasselt, weil ich zu viel erzählt hab' oder die Schüler zu viel gefragt haben. Aber ich glaub', bisher habe ich von keinem Kollegen deswegen bösen Kommentar gehört, ich meine, ich hab' auch gar nichts gesagt mehr zu denen, vielleicht sind sie deswegen stinkig gewesen" (L83, 7, 3–12).

[15] "Wenn man da jetzt irgendwo auf einen Punkt abheben will oder vertiefen würden, würden die anderen auch zur Verortung fehlen. Es ist halt doch Kontext" (L83, 21, 32–34).

Stage III—Teacher guides feel the need for recognition and further personal development

Since the observed teacher guides had worked for less than one year in their new position, they made no statements concerning long-term consequences for their self-image. In order to consider those effects which are relevant for Stage III, more interviewees were integrated into the sample: two teacher guides who have been working one day a week at the memorial for over nine years. One interviewee expressed how dissatisfied he was with the working conditions on site (L85, 1, 7). This teacher guide marks the transition to a new stage as follows:

> "The guided tours have been reduced from the original three hours to one and a half hours. As a result, the tours have become very standardized, because there's a certain amount of basic information you have to convey. It's difficult to bring in anything beyond that, things which perhaps also have to do with the area of commemorative culture, or perhaps an ethical emotional discussion. There is no institutionalized space for discussing such problems. You have to finish your tour within the three-hour time frame" (L84, 2, 36-42; translated from the German).[16]

While the teacher guides in Stage II consider 90 min for a tour appropriate, in Stage III, the time limits are regarded as challenges. The interviewee argues that on the one hand, a "culture of remembrance" takes time; on the other hand, an "ethical emotional confrontation" develops. These are important goals of touring which might conflict with schedules of school classes on site. Since he wants to involve the visitors with the historical facts and the history of remembrance, he wants to be free to plan the tour. In a similar way, teacher guides reflect on the category of space. They understand the site as a place of remembrance and not as a place to teach historical facts. Once more, the teacher guides expect freedom in creating a meaningful space. One teacher guide states that he creates "free spaces" within a guided tour:

> "It's good that we've always had the freedom to shape this guided tour. The training of [regular] guides at the memorial is much more narrowly focused. They work with a set of questions and with fixed stops on the tour. Of course, it's to our advantage that we as teachers can react quite flexibly to the groups. This is no longer possible when

---

[16] "Dass die Führungen von ursprünglich drei Stunden auf eineinhalb Stunden reduziert wurden, was zur Folge hat, dass sich die Führungen natürlich dann immer sehr stark standardisieren, weil man ein gewisses Grundmuster an Informationen einfach weitergeben muss und das darüber Hinausgehende, was vielleicht auch den Bereich auch Gedenkkultur, was vielleicht auch eine ethisch-emotionale Auseinandersetzung betrifft, das ist immer sehr problematisch, dann noch mit reinzubringen, weil man keine keine institutionalisierten Freiräume dafür, die man bei der dreistündigen Führung schon hat" (L84, 2, 36–42).

the level of standardization increases, which is as a result of the tight time schedule. This commemoration is one aspect, but I find that in the selection of topics or in the focusing of certain related topics, you actually hardly have an opportunity to address these in the 90 minutes" (L84, 3, 25–38; translated from the German).[17]

While in Stage II, the teacher guides reflect on how to create a perfect tour based on their own skills, the challenge in Stage III is to what extent the institutional framing and the personal ideals are regarded meaningful for their own goals. Above, the teacher guide expresses his desire to strengthen remembrance. At the same time, he thinks it's important to listen more to the other teacher guides and their individual experiences on site when planning tours and memorial events. The teacher guides express an awareness of a new professionalization, which one guide describes with an interesting comparison: "A famous soccer player from Hamburger SV, who couldn't perform, once said to his coach: Use me or sell me."[18]

The comparison reveals new aspects of interpreting tour guiding: The soccer player needs a team with a trainer. The idea is that there is some competition between the teammates to be successful as a team. This comparison can be interpreted as a form of individualized training and an increase in the variation of the tours at the memorial—depending on the groups' expectations. If the teacher guides' efforts succeed, then the game succeeds, the team is strong and can hold its own against an opponent—however it may be conceived: The "game" can be won when all members know what the game is all about ["was auf dem Spiel steht"]. Following this line of argumentation, the perspective of teacher guides is regarded as a "game" that has to be played.

In addition, the teacher guides want to contribute to a culture of remembrance, which helps them develop their new understanding of narrating history and integrate the visitors' desires. For example, the teacher guides find it important to establish a connection between Nazi crimes and the region from which the students come (L83, 2, 21). The role of the school during the Nazi period (L83, 2, 1 f.) also serves as a way to connect visitors' personal interest in the topic with the memorial site (L83,

---

[17] "Was nach wie vor gut ist, ist, dass wir schon eben schon Spielräume haben, diese Führung zu gestalten, also jetzt bei der Gedenkstätte, die Referentenausbildung ist viel enger geführt, die arbeitet mit Leitfragen, die arbeitet mit festen Stationen. Das ist natürlich unser Vorteil, dass wir als Lehrer doch ziemlich flexibel auf Orte und auch Personengruppen reagieren können ja. Aber das kommt eben durch diese hohe Standardisierung, die einfach durch diesen engen Zeittakt begründet ist, überhaupt nicht mehr zum Tragen, also dieses Gedenken ist die eine Seite, aber selbst find' ich in der Auswahl der Themen oder in der in der Fokussierung von bestimmten Themenkomplexen, hast du in den neunzig Minuten eigentlich kaum mehr eine Chance" (L84, 3, 25–38).

[18] "Ein berühmter Fußballer vom Hamburger SV, der nicht die Leistung bringen konnte, hat mal zu seinem Trainer gesagt: Setz' mich ein oder verkauf' mich" (L84, 3, 40–45).

2, 23 f.). This tendency seems to be reinforced in Stage III and has an impact on the narrative of the tours which the guides continue to develop. Compared to Stage II, teacher guides in Stage III seem to focus more on the students, with whom they want to familiarize the memorial. One teacher guide reports:

"For about two and a half years now when I meet the students and we introduce ourselves, I've actually gone over to questions like: Who are you? Where are you guys from? What type of school? Where are you in the curriculum now?" (L85, 1, 39–42; translated from the German).[19]

In Stage III, the guides with experience and training are more interested in individuality and subjectivity of their audience. They are in search of common ground between the visitors and themselves rather than providing information. For their long-term goal of developing a culture of remembrance, they keep in mind where the visitors, perpetrators, and victims came from. Thus, history of the Nazi period can also be experienced as local or family histories. Consequently, teacher guides contextualize history starting at the individual level and pointing to a more global perspective.

3   Summary of the Results

To sum up the different stages of teacher guides' development, I focus on specific questions which guides might ask themselves. In doing so, I describe the characteristics of tour guiding from an individual perspective.

- Stage I—Do I put myself at the service of the information and intensification structure of the space? Do I succeed in dealing with the subject in such a way that I pass as a teacher guide? Can I connect the cognitive body of knowledge with the silencing space? What does it mean to work at a memorial site?
- Stage II—How can I optimize my tour so that it does justice to the changing audience and so that I myself, as a pedagogically skilled person, am satisfied with my performance?
- Stage III—What are the limits of the propositional approach to the space and its history? Does the space demand more than a cautionary walk-through, that is, something that could initially be described as commemorative? Do I meet

---

[19] "Ich bin tatsächlich seit ungefähr zweieinhalb Jahren dazu übergegangen, meine erste Frage, wenn man sich begrüßt und vorstellt, zu fragen: Wer seid ihr? Wo kommt ihr her? Welche Schulform? Wo seid ihr jetzt im Stoff?" (L85, 1, 39–42).

the students as individuals? Can I commemorate together with them and with the teacher accompanying them?

My goal in this paper was to show the dynamic process of self-reflection among the teacher guides at the memorial site in Dachau. Two main results have emerged and are worth mentioning: First, I observed how self-awareness of teacher guides develops in a process of three stages which is based on the background of their teaching profession. Second, it is striking that in the third stage, a need for commemorating and slowing down the process of mediation becomes apparent. Teacher guides are not satisfied with the idea of touring as an optimization of touring. They continue to ask questions and struggle for individual perspectives. They want to start a conversation with visitors and develop new forms of encounter at the memorial site.

It should be emphasized once more that the desire of the teacher guides for a strengthening of remembrance are tantamount to an appeal to those involved in Holocaust education. For example, artistic forms of dealing with the Holocaust could support these efforts. In this context, the projects reflected by Heise can be mentioned (Heise 2015). Religious education approaches could also be useful and can build on the suggestions made by Danner for small and often unknown memorial sites in Austria (Danner 2020, p. 258–265; Danner in this volume). In a positive sense, the forms of memory can build on the "remarkable diversity" that Stevick (2017, p. 191) found among teachers who teach the Holocaust—and that Gudehus already found important in his call for "more individual and plural narratives" about the Holocaust (Gudehus 2006, p. 238). Currently, it is important to discuss teacher guides' personal experiences, and support them in actively encouraging conversation with visitors. This also includes that teacher guides start a dialogue with memorial workers on site. All memorial workers on site should be obliged to emphasize human rights in light of Holocaust education.

References

Ballis, A. (2019). Guides an KZ-Gedenkstätten und Holocaust Museen—Professionalisierung in Zeiten eines Wandels der Erinnerungskultur. In A. Ballis, & M. Gloe (eds.), *Holocaust Education Revisited. Wahrnehmung und Vermittlung—Fiktion und Fakten—Medialität und Digitalität* (141–166). Wiesbaden: Springer.
Danner, S. (2020). *Niemals Nummer—Immer Mensch. Erinnerungslernen im Religionsunterricht*. Göttingen: Vandenhoeck & Ruprecht.
Gudehus, C. (2006). *Dem Gedächtnis zuhören. Erzählungen über NS-Verbrechen und ihre Repräsentation in deutschen Gedenkstätten*. Essen: Klartext.

Hammermann, G., & Pilzweger-Steiner, S. (eds.) (2017). *KZ-Gedenkstätte Dachau. Ein Rundgang.* Dachau: Stiftung Bayerische Gedenkstätten.

Haug, V. (2015). *Am "authentischen" Ort. Paradoxien der Gedenkstättenpädagogik.* Berlin: Metropol.

Heine, M. (2019). *Verbrannte Wörter. Wo wir noch reden wie die Nazis—und wo nicht.* Berlin: Duden.

Heise, J. (2015). *Über die Sprache hinaus—künstlerische Vermittlung an Gedenkstätten.* In E. Gryglewski, V. Haug, G. Kößler et al. (eds.), *Gedenkstättenpädagogik. Kontexte, Theorie und Praxis der Bildungsarbeit zu NS-Verbrechen* (302–318). Berlin: Metropol.

Simms, B. (2019). *Hitler. Eine globale Biographie.* München: DVA.

Stevick, D. (2017). Teaching the Holocaust. In M. Eckmann, D. Stevick, & J. Ambrosewciz-Jacobs (eds.), *Research in teaching and learning about the Holocaust. A dialogue beyond borders* (191–222). Berlin: Metropol.

Links

www.dachauer-forum.de/themen/gesellschaft-geschichte/guided-tours-dachau-concentra tion-camp-memorial-site/. Accessed 16 December 2020.

www.blz.bayern.de/politische-bildungsarbeit.html. Accessed 16 December 2020.

Teachers as Guides at Memorials and Places of Remembrance

Sonja Danner

> *"Ich komme nicht von Auschwitz her—ich stamm' aus Wien."*
>
> *"I don't come from Auschwitz—I come from Vienna."*
>
> Ruth Klüger

Abstract

At smaller places of remembrance and memorial sites, teachers play an important role since there are no official tour guides. Focusing on such visits to memorial sites in Austria, teachers were interviewed to understand their practice of guiding at those places. Using a qualitative approach, all interviewed teachers agree that student participation in the excursion must be voluntary. Especially in the subjects of religion and art education, teachers stress a holistic approach. Further, they recognize the importance of dealing with one's own family history, both for themselves and for the students. Biographical work is the pivotal point of the field trips. This is intended to encourage a change of perspective and empathy. Preparation and follow-up work are recognized as essential elements of memorial site visits.

S. Danner (✉)
Campus Wien-Gersthof, Kirchliche Pädagogische Hochschule Wien/Krems, Vienna, Austria
e-mail: sonja.danner@kphvie.ac.at

© The Author(s), under exclusive license to Springer Fachmedien Wiesbaden Gmbh, part of Springer Nature 2022
A. Ballis (ed.), *Tour Guides at Memorial Sites and Holocaust Museums,*
Holocaust Education – Historisches Lernen – Menschenrechtsbildung,
https://doi.org/10.1007/978-3-658-35818-1_6

1   Introduction

Since the mid-1980s, the Ministry of Education has strongly recommended that school classes visit the Mauthausen Memorial in Austria. Teachers of history, social studies, and political education, as well as those who teach German, often comply with this recommendation and organize class trips. The fourth grade of the lower secondary school (14 years of age) and the seventh grade of the upper secondary school (17 years of age) are particularly suitable for this purpose, since the Second World War is also part of the curriculum in both grades. Through these excursions, "the Holocaust" is increasingly located in Mauthausen, near the Upper Austrian capital Linz. This often gives students the impression that mass killings are limited to the former concentration camp Mauthausen and that the mass exterminations have little to do with their own local center of life. "Evil" can thus be easily projected onto "the others."

It is different in religious education (RE) courses. They are often characterized by small group sizes in schools, which sometimes make it difficult to organize a trip to Mauthausen (due to costs, students coming from different classes, and other reasons). In order to avoid these difficulties and still be able to work with young people on the topic of the Shoah at historical sites, teachers of RE all over Austria include small places of remembrance into their courses, which are located near the school or the students' homes.

In both cases, instruction occurs in a dislocated manner— outside of the school—and yet the memorial and the place of remembrance already differ substantially from each other in the conceptualization of teaching. While the term "memorial pedagogy" [Gedenkstättenpädagogik] gives the impression that there is only one approach for all the memorial sites, a pedagogical approach for places of remembrance is completely missing. Memorial pedagogy is used when we are talking about places that bear witness to the witnesses beyond their death and which—together with architectural monuments or memorials and museums—form an overall arrangement. In recent years, pedagogical approaches and suggestions for memorial sites have become more and more popular in German speaking countries. While memorial sites provide pedagogical guidelines and also guided tours, places of remembrance are usually not direct crime scenes and are free of any pedagogical conceptualization. Here teachers often use a general pedagogical strategy that is independent of sites. This is called memory guided learning [erinnerungsgeleitetes Lernen] (Danner 2020, p. 114 f.).

The historian Zülsdorf-Kersting draws attention to the fact that memorial sites are losing their significance as places that are accessible to visitors through their personal memories, and that they must be seen in a new light. If they are taken as places of learning, the necessary learning arrangements must be made available, and the history of National Socialism must be conveyed. They are therefore places of intentionality, with the aim of not only enlightening the visitors and conveying certain contents and contexts, but also influencing their views and attitudes (Zülsdorf-Kersting 2011, p. 172). However, these intentions are not free from "Zeitgeist," that is, they are influenced by the work of various forces such as politics and business. Teachers are therefore requested not to adopt the given pedagogical guidelines without reflection, but to address them in class. They have to deal with expectations of educational policy control and optimization because, as the historian Meseth assumes, behind many reforms there is a rather technological understanding of education (Meseth 2015, p. 99). Meseth also raises the question of the orientation of teachers. Do they follow an output-oriented approach[1] or educational theory approaches? In the latter case, they do not arrange their teaching around a certain strategy or method, but instead work on a case-by-case basis to find good solutions. Meseth compares this with the approaches of a lawyer, doctor, or pastor (Meseth 2015, p. 106 f.). Thus, teachers should first think about what their intentions are and which goals they want to achieve in their lessons with different students, as Young summarizes: "Where does all the history and its telling lead, to what kinds of knowledge, to what ends?" (Young 2004, p. 22).

Lehrke, the former head of the office for cultural activities in Bremerhaven (Germany), also finds that a specific feature of memorials as places of learning is that learning at these historical sites is limited in time (Lehrke 1988, p. 28). However, not only does the public (or better: official) culture of remembrance guide the experience, individual constructions of history by the students themselves, usually strongly influenced by their own family narratives, must also serve as a basis for further educational considerations.

Since there are no ready-made pedagogical guidelines for teaching groups at places of remembrance, and memorials also differ in their pedagogical guidelines in Austria, the following is a brief outline of the basic factors for the pedagogical and didactic actions of teachers.

---

[1] They are mainly focused on factual knowledge which students should be able to reproduce.

2   State of the Art—Key Incidents for Teaching at Memorials and Places of Remembrance

The central point of any preliminary consideration for teaching at an out-of-school learning venue—either in a place of remembrance[2] or at a guided tour in a memorial site—is the subject, the students. It has to be clear that the visits for them are voluntary. In religious education studies emphasis is placed on a "double-subject-orientation" (Boschki 2001, p. 362), which, on the one hand, focuses on the students as learners (the present) and, on the other hand, on the victims, spectators, and perpetrators (the past). Students thus not only become addressees of an educational program, but also see themselves as part of history. Education therefore becomes self-education (Wagensommer 2009, p. 68 f.). This approach, however, does not take into account the teachers, who are supposed to initiate the learning process from subject to subject and play an essential role in the teaching process. Since the publication of numerous books on the Hattie Study (Hattie and Zierer 2018) in German speaking countries, the significant effects of teacher personality on the learning success of students in education have been highlighted. This has to be taken into account here as well, and the double-subject-orientation needs to be supplemented by yet another subject: teachers. Thus, at least two of these subjects are anchored in the present with a view to sustainability (and thus to shaping the future). Victims, bystanders, and perpetrators, on the other hand, are less and less likely to be found in the present, since contemporary eyewitnesses are either no longer available or have passed away, due to their advanced age. For this reason, the synchronous confrontation will gradually give way to focusing on archives and existing material about the eyewitnesses. Even projects like 3D "holograms" from the USC Shoah Foundation[3] will not change this, although they sometimes give the viewer the feeling that the person is still alive (Ballis in this volume).

The German theologian Fuchs supplements the concept of double subject-orientation with the concept of encounter, which to him seems to be central in the process of memory learning (Fuchs 2001, p. 309). However, subject-orientation also means thinking about the personality of the subjects and their history or histories. These identities, influenced by many different factors, are not a "tabula rasa." Each one entails preconceived opinions and approaches to

---

[2] Talking about a guided tour here means accompanying the students in coming to terms with history at the site, since places of remembrance are often limited to a memorial plaque or a memorial stone.

[3] www.yadvashem.org/de/education/newsletter/10/holograms-and-remembrance.html.

the subject, which are influenced by micro-social factors (family) and factors within the macro-system (environment, media, etc.). Here, too, research shows that visits to memorial sites tend to reinforce and cement existing attitudes rather than soften and change them. Self-reflexive processes are hardly ever initiated; acquired knowledge refers mainly to the site itself but fails to establish historical connections (Pampel 2011, p. 20–22). From the perspective of history education, Zülsdorf-Kersting, a professor of didactics of history, therefore calls for teaching and learning arrangements to set in motion individual and historical thought processes for reflection. The aims of the learning processes are not to encourage superficial expertise of spontaneous reactions without reflection at the memorials themselves, but rather to integrate the new impulses into solidified images of history. Students should be encouraged by their teachers to analyze facts self-reflexively and to judge facts and values (Zülsdorf-Kersting 2011, p. 190 f.). In this context, the relevance of the historical events for the present and for every human being is often mentioned in the specialist literature for teachers. This does not mean making comparisons between then and now, but rather developing an awareness of history. This can be seen "in the effort to achieve empirical reliability, balancing judgments and reflect[ing] on evaluations and perspectives" (Kaiser and Rinke 2015, p. 148).

To underline the importance of teaching, that the Holocaust is still relevant for the present, Herrmann, who studied sociology, politics, and education, points out that when it comes to relating to the present, a consciousness of time must be created by the students. Outer/inner time or world time/own time must be synchronized. The personal perception of time often does not correspond with the real/world time. Only when this time awareness is present can a bridge be built to a reflected future after Auschwitz. Remembrance work always moves on these different time structures and lives from the connections that are recognized by the individuals. When these connections are recognized, personal consequences can be drawn from the confrontation with the time of National Socialist governance (Herrmann 2007, p. 25). When education of the individual is seen in this light, it refers to the ability "to choose times, to be able to withdraw from times in order to take time for processes of reflection." Herrmann further concludes that educational processes result "between time heteronomy and time autonomy" and sees the acquired competence to act in a critical awareness (Herrmann 2007, p. 29). In contrast, Meseth critically questions the competence model in relation to educational practice when it comes to this particular way of communicating National Socialism, as it falls short of what is required. As valid justifications, he proposes the moral implications of the topic and the specific structure of this out-of-school learning location (Meseth 2015, p. 103).

Again and again it has been pointed out that students in general, and especially during visits to memorial sites, have to be met where they are. For Sternfeld, this notion does not go far enough. The cultural historian stresses a different position: She emphasizes releasing young people from where they currently are in order to empower them to think and act in a self-determined way (Sternfeld 2003, p. 98 f.). This presupposes, however, that their teachers will help them find their way. To ensure this, teachers must be able to reflect on their own family histories and on personal approaches (rational and emotional) to the topic in general. Only in this way is it possible to deal with a controversial topic such as the Shoah at memorials, and to go along with students and strengthen their ethical and self-critical orientation and judgment[4] without taking a moralizing approach. This is extremely important for a reflective approach to the topic especially in Austria, where the victim myth persisted until the 1980s, memorials and places of remembrance have been repeatedly vandalized, such as Luigi Toscano's exhibition *Against Forgetting* in Vienna. Portrait photos of victims of National Socialist persecution were cut up in May 2019.[5] Such vandalism did not occur at any other place where the exhibition was shown. There was only a very short public debate, and the trespassers have never been found. Public debates on the subject of the Nazi period are rarely conducted. Possibly there is a connection here with silence on the topic in families who avoid discussing this time period because they often do not want to know about their forefathers' involvement in Nazi crimes. Those who do not speak about this explosive topic within their families, also do not talk about it in public. Rosenthal has dealt with silence within the generations and found that in families of both perpetrators and persecuted persons, these secrets have an immense effect. They are among the most effective mechanisms when it comes to the latent persistence of a family's past—the more that is concealed, the more lasting the impact on the generations of children and grandchildren. Not talking reciprocally hinders not only the discussion of the past but also the adoption of other perspectives through lack of dialogue (Rosenthal 1997, p. 19–22).

These voids, which thus arise in the biographies of the following generations, then sometimes provoke "search movements, the formation of myths and

---

[4] The theologian Boschki mentions judgment as one of seven essential guidelines for memory-based learning in the classroom. He describes the others as follows: Ethical and self-critical orientation (judgement), double-subject-orientation, Religious Education as Human Rights Learning, biographical and local orientation, the larger context of Jewish-Christian learning, prohibition against overwhelming students, and combating antisemitism (Boschki 2015).

[5] www.diepresse.com/5651705/fotograf-luigi-toscano-allein-und-ziemlich-fassungslos.

the development of counter-identities, serious concerns about family involvement in National Socialism;" some people fill these gaps with fantasies. Likewise, emotional and cognitive approaches to the topic of National Socialism depend strongly on family backgrounds and influence (Wagensommer 2009, p. 356). Biographical work is therefore an essential key in the didactic considerations for memorial visits. Both the biographies of teachers and students, and those of the "subjects to be researched" play a major role. Teachers are especially challenged to encourage their students to enter into dialogue with their own family members, to explore family narratives, and to research family history. The question—Where do I come from?—is essential in dealing with the history of National Socialism and also the question—What does this have to do with me? Consequently, the two areas—the historical and the family approach—do not remain separate but are brought into dialogue. Dialogue does not mean alternating speech or exchange of information, but rather active discussion on different levels: confrontation with oneself (one's own approaches, questions, religious search processes), with the fate of others (commemoration of the victims), and with the society (political church reality in the past and present) surrounding us. A dialogue-driven approach means working on questions rather than providing ready-made answers (Boschki 2001, p. 368). Meseth sees this aspect anchored in one of his three horizons of interpretation, namely that of remembrance. Here he poses the question of who commemorates how (individually, collectively) and how the individual relates to the victims of National Socialism. As further horizons of interpretation, he mentions a topic and site-specific expertise and a normative-ethical perspective, which is intended to sharpen human sense and sensitivity by dealing with what is inhumane (Meseth 2015, p. 109).

Krondorfer, a professor of religion at Northern Arizona University, is in favor of an active shaping of possibilities for transformation and confrontation in memory work. In order to understand one's own dreams and memories, an analytical, associative and historical interpretation is required. The shadow sides preserved and expressed there require awareness of their affective effects; this awareness releases individual and collective anxieties and can lead to regressive behavior of the collective ego. If this is not faced, people can get stuck in their own family narratives, avoidance or victim discourses, and this can manifest over generations (Krondorfer 2013, p. 486).

Fundamental for any kind of biographical work is the ability to empathize and change perspectives. Empathy here means neither a distanced, purely emotional identification, such as sympathy, nor an abstract, cognitive understanding of a person and his/her emotions. Rather, it is about reducing distance to the historical person "through empathetic imagination" and later restoring this distance in the form of a "narrative" as a reflective act on one's own person (Schellenberg 2015, p. 133). Distance is indispensable in order not to identify oneself completely with the other person but to understand and judge one's own personality (thinking, feeling, acting), because the ability to act is only possible from a certain distance. This also means that it is not only victims who must be brought into view, but also perpetrators, who must be the subject of biographical work, in order to be able to reflect on the perpetrator in oneself. In a protective setting, the change of perspective should offer the possibility to reflect and question one's own points of view without running the risk of being immediately attributed to a group or not. It is much more a search movement that can have an identity-forming character for students and their classes/groups.

Within a biographical approach a protective setting is essential, especially for multicultural learning groups, where students bring not only their own cultures of remembrance, which—influenced culturally and shaped by religion—can be different from one country to another. Consequently, young people's biographies differ. Frölich from the Anne-Frank-Center Berlin points out that especially in such groups, "a self-reflexive dealing with personal identity constructions as well as the interlocking with structures of dominance and subordination" is indispensable (Frölich 2010, p. 136). To cope with these challenges, well-prepared learning encounter groups and sensitive communication are recommended. By these means, individuals are able to open themselves to this process without the risk of being hurt.[6]

An out-of-school place of learning can be very conducive to open up as the setting of teaching outside the familiar school building will be different from the classroom setting. The two geographers Sauerborn and Brühne here see the opportunity to a questioning and researching unlocking reality by active appropriation (Sauerborn and Brühne 2012, p. 84). They also warn that the didactic demands on out-of-school learning can only be covered to some extent, because of the range of variation: explorative learning via experience-based and student-oriented approaches, various methods and different social and action forms (Sauerborn and Brühne 2012, p. 26 f.). However, the action-oriented approach to out-of-school places of learning also has the potential to promote learning and motivation and to awaken students' own initiative. In order to exclude random learning and avoid arbitrariness didactics, it is important to relate the results of research-based learning to superordinate subject areas (Sauerborn and Brühne

---

[6] This conclusion emerges from the evaluation of student feedback in Holocaust Education classes from 2014 to 2020 at the University of Vienna and at the private University College of Vienna run by Sonja Danner and Christian Matzka.

2012, p. 28). Sufficient learning material must be made available, so that the demand that teachers should control and guide out-of-school learning as little as possible (Sauerborn and Brühne 2012, p. 71) can be easily implemented. These ideas correspond with the results of a survey of teachers and students, according to which students should be encouraged to form their own opinions through active, explorative, research- and project-oriented learning. The importance of getting to know the biographies of victims and perpetrators is also emphasized (Pollak 2011, p. 247). The theologian Schwendemann here speaks of a participatory pedagogy which must find its way between the single memorial or place of remembrance and a communicative-participative transfer structure (Schwendemann 2013, p. 109). Since historicization is also advancing in this teaching topic, Ritscher as a psychologist, views learning at memorial sites as being primarily based on relationship and communication within the setting of a process-oriented and personality-building learning (Ritscher 2013, p. 210 ff.). Historicizing is the process by which subjects of current events become subjects of historical interest. It has been a central method in humanities and social studies for a long time.

To cope with this multitude of requirements and to be able to organize extracurricular lessons effectively, both teachers and students must be well prepared. Every place can hold potential for conflict and is often characterized by an emotional depth. These factors cannot be planned but must be addressed by the teacher. For Meseth, "therefore not only methodological and didactic know-how is required, but also a broad technical and pedagogical knowledge. But above all a knowledge of the historical needs of young people, which can be expressed in a special way in this subject." (Meseth 2015, p. 100; translated from the German). In addition to the preparation for a visit to a memorial or a place of remembrance, follow-up work must also be carried out so that students are not left alone with their cognitive and emotional reflections.

These are some of the theoretical inputs from different points of view within the last years. In the following, we will examine classroom practices in some detail.

3 Teachers and Their Educational Considerations—An Empirical Study at Austrian Memorials and Places of Remembrance

In recent years, research in the field of memorial pedagogy as well as in history didactics and religious pedagogy has primarily been driven by the desire to learn more about the effectiveness of memorial site visits by students. Teachers and their teaching activities were not the focus of attention.[7] One of the first studies from 2020 focusing on the work of teachers is a research study about Protestant religious education teachers at places of remembrance in Austria (Danner 2020). This article summarizes the results of a comparative study that deals with different approaches by teachers to guided tours at memorials and places of remembrance.[8] A total of 17[9] semi-structural, problem-oriented, guideline-based interviews were conducted from 2012 to 2016, ten of them with teachers of Protestant religious education (RE)[10] (2015/16) and six with teachers of history, social studies, and political education (GSP), as well as German (D) (2012).[11] Another interview was conducted with an art teacher at a school that focuses on fashion, art, and design (2012).[12] The teachers were asked about their motivations for organizing field trips with students and about their didactic preparation and their role as teachers/guides at the memorial itself. Included are teacher surveys from three different projects.[13] In their statements, the teachers indicate their desire to include memorial sites or places of remembrance when teaching about the Shoah. The interviews were transcribed according to the minimum standards for transcribing interviews for the archive of the Mauthausen concentration camp

---

[7] A comprehensive overview of studies on Holocaust education can be found in Danner (2020, p. 15–41).

[8] The term memorial is used for places that have pedagogical guidelines and are often connected with an exhibition or a museum. Place of remembrance, on the other hand, is the term used for a place that is not associated with an exhibition or museum and is often a silent witness—mostly unknown—to the Holocaust. That can be a commemorative plaque or so called "Stolpersteine." These are brass plates embedded in the street to remember the fate of people deported, persecuted, and murdered by members of the Nazi Regime.

[9] At the time of the interviews, more than half of the teachers were between fifty and sixty years old. Significantely more men than women (70%) volunteered to participate in the interviews, although the distribution of whether they taught at lower secondary level (10–14 years) or upper secondary level (14–18 years) was roughly balanced.

[10] They are quoted below with the capital letters A-J.

[11] They are quoted below with the capital letters K-P.

[12] He is quoted below with the capital letter Q.

[13] In a project from 2012 to 2014, teachers who brought their students for a guided tour to the memorial of Mauthausen were interviewed. In another project in 2012, an art teacher was interviewed. He did the guided tour with his students in Mauthausen as well, but, in addition, asked the students to document their own important discoveries at the site with cameras. Last but not least, RE teachers who went to small places of remembrance in different regions in Austria were interviewed from 2015 to 2016. They visited places of remembrance in their hometowns that commemorate events of the Holocaust, e.g., places where synagogues were destroyed and from where people were deported. All interviews were conducted in German language and for this paper translated into English.

memorial site. The evaluation was based on the qualitative content analysis by Mayring (2015).

While the six teachers of history, social studies, and political education (GSP) and German (D) took advantage of the classic offer of a guided tour in Mauthausen and travelled to the memorial site by bus, RE teachers went mostly by foot with their groups, as they visited small places of remembrance near their schools.[14] Some of them also went to Mauthausen when they made themselves available as accompanying teachers for field trips with the whole class (not only with their religious group), but usually did not have a leadership role. The art teacher took a middle course, because he planned to go to Mauthausen with a two-hour guided tour and to visit other smaller places of remembrance with the same group of students.

The main consideration regarding the field trip was to decide whether to include a large memorial or a small place of remembrance in the lessons. In Mauthausen, pedagogical guidelines were developed, and trained guides were happy to offer guided tours. The teachers who were interviewed assumed that their role at the memorial was limited to looking after the "sensitivities"[15] of individual students and answering any additional questions. The experience of K shows that the students preferred to address their concerns to the person they were familiar with and not to the official guides. In this case, teachers were more like co-guides and rarely decided on the content. D even stated that he deliberately declines guided tours in memorials with an existing program, because he knows his students much better than the guides do and can, therefore, respond to them better. RE teachers usually cannot access any program, since small places of remembrance are only in some cases prepared for teaching. As already mentioned, it is difficult and expensive for small RE groups to travel to memorials like Mauthausen; it is also inconvenient to take some students out of their classes for the trip while others have to follow the lessons provided in the school schedule. Therefore, RE teachers have to create their own pedagogical goals for the sites, which has the advantage that they can be adapted to the group of students. However, this means additional effort in the lesson planning for the excursion. What motivates all of these teachers to conduct a visit in the first place, and how do they prepare their students for the trip? How do teachers plan their lessons at

---

[14] The religious groups usually consist of less than ten students from different classes and grades. Thus, a visit to Mauthausen is hardly feasible in terms of organization and finances.

[15] Sometimes it happens that students are overwhelmed by strong feelings and need somebody to talk to.

the memorial or the place of remembrance and how do they follow up? This will be described in more detail in the following.

3.1   Teachers' Motivation

Q sees his task as a teacher in pointing out connections. He wants to help convey a differentiated and ethically justifiable view of the world. In doing so, he has in mind the discussions about "visual communication" in art education, with his focus on photography. The aim is to reflect, analyze, and verbalize media products. Students gain knowledge by producing posters, logos, and photo series.

Teachers of the subjects GSP and D feel a moral obligation in teaching the subject and want to present history in an appropriate way. The memorial plays an important role, since it has an effect per se and makes history more vividly perceptible. Therefore, everyone should go to Mauthausen once, without being morally pressured to do so. While the aim of the visit is to provide students with arguments against Holocaust deniers and to create an awareness (L) of what human beings are capable of, teachers should not make students feel personally responsible for the events of the Holocaust.

RE teachers also visit places of remembrance because they expect a special impact at the out-of-school learning place. The aim is to bring out emotions, meaning that students are touched or affected, which should reduce the historical distance (F). The proximity of the place to the school or the place of residence plays an important role, because the place of remembrance is sometimes familiar to the young people, and they can relate to it more easily.[16] It makes them aware that the Holocaust is not only located in Mauthausen, but also in their own surroundings, and it is linked to the real fates of individuals (A). However, the place of remembrance should not only have an emotional impact, but should also provide knowledge. These two motivations are important conditions for transferring the history into the presence of the students and for creating an awareness of one's own responsibility and duty to society. "This has something to do with us here and now," as F states. Finally, the culture of remembrance has to be mentioned, which plays an important role in RE. The churches must not forget their complicity (B). The motives of the RE teachers to plan field trips go hand in hand with the following objectives.

---

[16] Proximity not only describes the fact that the place of remembrance is close to students' homes or schools but is also intended to reduce the historical distance through the personal relationship of the students to the place.

3.2   Objectives

For Q, action-oriented learning is important, which means practical action and "visible verification of individual imprints and values in the form of creative work." It is important for him (Q) to emphasize that he can consciously follow his own will and thus be able to act freely. Teachers of GSP and D focus on competence orientation, and sometimes they have very ambitious objectives. However, they also want to promote the students' own experience of "seeing the place as it really existed and feeling what happened there" (L). The resulting personal consternation, also sometimes shaped positively, should be reflected, which might lead to personal actions, resistance, and civil courage according to the motto: I don't want to be determined by others. (K). Also, personal actions—especially in the class—should be questioned. A "wake-up call" to students with right-wing extremist attitudes is being considered by some of the teachers, a "vaccination against the right-wing." How this transfer from knowledge to personal action (stand up for right-wing extremism) can be achieved is not mentioned in the interviews. Teachers rely on the multiplication function of their students for peers and for the parental home when it comes to developing an awareness of right-wing tendencies in society. However, the teachers also express their fears that the process is not verifiable and that it is therefore not possible to trace the failure to achieve to the learning objectives. In religious education, teachers do not only focus on historical facts[17] but also on ethical and psychological knowledge[18] and set their learning targets in a similar way to teachers of other subjects. They want to clarify the role of history in students' everyday lives, preventing repetition of crimes and cruelties. The responsibility of individual persons for the present and future in society is reflected in the hope of preventing right-wing populist ideas. In order to come closer to these learning targets, teachers stress the necessity of an extensive discussion in the classroom before visiting the memorial sites.

---

[17] They are very interested in the history of the Protestant church, connected with the respective place and also in the fact that this knowledge transfer creates complements to other subjects. For this reason, prior arrangements are made with teachers of GSP, D, and Catholic RE. These colleagues often cooperate with each other in the classroom in an interdisciplinary way. Teachers of GSP and D, on the other hand, mention that there is hardly any cross-curricular teaching, unless one teacher has several subjects and coordinates the material across the subjects. Other subjects are only in some cases worked with by dedicated teachers.

[18] The mechanisms such as the devaluation of people or the deliberate use of violence and its legitimization are mentioned (E).

3.3   Preparation for the Field Trip

For Q, thorough preparation for a visit to a memorial site and a follow-up discussion are critical aspects of the learning experience. This also applies to RE
teachers, who attach great importance to a sound—especially cognitive—preparation for their students and for themselves. The essential question for them is: "Can
I manage that?" (I), without overwhelming both the students and themselves.
Therefore, these teachers prepare very well by reading books or subscribing to
relevant journals, and they often have their own library on the subject. The Internet is of less importance as a source of information. Some also visit the place of
remembrance in advance. Only one teacher prepares herself emotionally (C) and
takes this experience as impulse for her lessons.[19] For teachers of the other subjects, neither their own preparation nor that of the young people is an important
issue.[20]

Nevertheless, two teachers are very concerned about the students and discuss
the visit to the memorial site in Mauthausen in advance with the young people (K, P). They are involved as persons with their own vulnerability and fears,
with their personal family narratives and uncertainties[21] and consider both cognitive (information about Mauthausen) and emotional preparation of the students
(addressing fears and expectations) to be essential in order to prevent displace-
ment activity.[22] In group work, projects, and individual research topics such as
terror and persecution, as well as Austria in the National Socialism Era, are prepared and presented to the class. In doing so, the teachers also reflect on including
their own and students' family experiences. RE teachers, however, have their own
family background in mind, but do not encourage their students to explore theirs.
Instead, the approach is strongly influenced by knowledge transfer and is only
somewhat complemented by emotional and action-oriented approaches.

---

[19] This teacher is one of two who try to prepare the students holistically. The students get historical knowledge and they are prepared emotionally, by talking about emotions that could
emerge during the visit. Texts and letters are also written by the students themselves to
raise awareness of the place and to encourage questions. Only once a spiritual impulse is
mentioned, which should distinguish the RE lesson from other subjects.

[20] One person (M) indicates that she simply waits for questions from the students in class.
She obviously assumes that she can answer them readily and that she does not have to prepare
herself in advance.

[21] This approach turns out to be the basic attitude of RE teachers.

[22] In one case (F) the teachers describe that a student who was completely overwhelmed
exhibited a childish conspicuous behavior. At worst, students can build up resistance to the
topic and block it out (Schwendemann 2013).

One argument of GSP and D teachers for not preparing their students as thoroughly is that there is too little time for deeper discussions and that the class sizes (up to 28) make extra preparation difficult for teachers. Other arguments are also mentioned which prevent preparation: lack of funding, not enough eyewitnesses to the Holocaust, and lack of opportunities for cross-curricular teaching. Nevertheless, these teachers name several media they find helpful. They use the documents provided by Mauthausen or found on the homepage of the memorial site. Further, they use newspapers and relevant films, for example, *Der Bockerer*, *Schindler's List*, and the *Mühlviertler Hasenjagd*, as well as documentaries such as *Austria Our Century*, without discussing the diversity of sources and their meaning with the students.[23] In this the two groups do not differ. RE teachers also use films as sources. *Anne Frank*, *Napola*, and the *White Rose*, for example, are used to quickly get the students acquainted with the subject matter. Mostly, however, RE teachers use documentaries, for example, eyewitness reports and biographies, as an introduction to biographical learning that helps to establish a connection to the place. In addition to books like *Damals war es Friedrich* or *Anne Frank*, they draw increasingly on Christian texts (*Zeit der Umkehr. Die evangelischen Kirchen in Österreich und die Juden*, or *Are We Still Useful?* by Dietrich Bonhoeffer). The aim is to use these media to focus on everyday life under fascism and the role of the Protestant Church under National Socialism. Although the reasons for visiting memorial sites are manifold, all interviewed teachers agree that the visit must be on a voluntary basis.

3.4 Teaching at Memorials or Places of Remembrance

The educational approaches to the visits of memorials and places of remembrance are different, as the local conditions are very different.[24] GSP and D teachers took two different guided tours in Mauthausen. In one tour, the students were accompanied in small groups through the grounds; in the other, a guide accompanied the whole class. The role of the teachers was mainly to provide the guides with information about the group and to determine together with them the main focus of the memorial visit. It was important for teachers that the visit to Mauthausen provided information on individual fates as a supplement to the

---

[23] There is also no reference to source criticism among RE teachers.

[24] On the one hand, out-of-school places without pedagogical guidelines and hardly recognizable as places of remembrance are visited, but on the other hand, well-prepared memorial sites are also visited. In addition, for RE teachers, persons (eyewitnesses) are also considered places of remembrance (Marko Feingold†). They are also included in the lessons.

factual knowledge gained in class. They also believed that the emerging emotions of the students at the memorial should be discussed to deepen reflection on the topic. In large groups, the teachers' support was helpful in this respect to avoid leaving the young people alone.[25] Special mention was made about the different ways in which girls and boys dealt with their concerns during the visit. Boys were said to have fewer articulation skills. A fixation on visiting the gas chamber was definitely not desired by teachers. Instead, teachers and guides wanted to get students to ask questions.[26] Q also accompanied his group with a guide on the joint tour. Afterwards, the students had another two hours to walk through the area with the camera alone or in small groups. Notes of their thoughts and emotions were also written down. The teacher was always available for questions and supported wherever help was needed. Q justifies this by attributing greater learning success to the memorial visit among the young people if the emotions that arose can be expressed. This must be included in the preparation. He is very skeptical about the mere imparting of knowledge, since a lot of knowledge is also available in the right-wing extremist scene, but no development of empathy or ethical conclusions can be seen from it. Knowledge alone is no guarantee for change in a society.

Within the sample, the role of RE teachers is completely different. Not only do they adapt the visits to the group of students, they also act as guides themselves, and they are always available for their students as a projection screen and contact person. They often have close bonds with their students, many of whom they have had in their classes over several years. If small memorial sites with available educational material (films etc.) are visited, this is optionally included and provided with work assignments. In this way, factual learning and ethical learning are combined. Biographical work[27] and the teacher's narratives are at the forefront of their educational considerations. A moment taken for self-reflection marks the end of the visit. Since the teachers are mostly second post-war generation, they have the function of bringing the past to the present, sometimes by telling their own family stories. In some cases, texts of eyewitnesses are read aloud. This educational approach is also chosen for visiting small places of remembrance that

---

[25] The reason given for this was that students have many images of the Holocaust in their heads due to fragmentary knowledge. That needed to be processed. The teachers also pointed out the important additional support of the parental home, especially when it comes to helping young people who are emotionally affected at the memorial, to better categorize their feelings and factual knowledge after the visit.

[26] This is also a major concern of RE teachers.

[27] Three teachers pointed out that it is important to them not to leave the victims anonymous, but to put a face on these people by naming them and telling their stories.

are located outdoors. One third of the teachers even put together a route to visit several small places of remembrance during a single excursion.[28] For example, an RE teacher walks in the footsteps of the death march of Hungarian Jews, and the young people have the opportunity to reflect here, among other things, whether it was really possible that nobody saw this group of people, although the route passes close to populated areas. Another teacher deliberately selects places of remembrance, which are well known to the students through their youth clubs in the city center. Here, as a result of the visit, the young people's view at familiar places of the city center changes. The local conditions are also included in the didactic considerations. Thus, the cross shape of a place of remembrance—almost predestined for RE classes—is used at various points to develop empathy and is explored and reflected upon with elements of self-awareness. A further third of the RE teachers already have previous experience in helping to design a memorial and incorporate this into their lessons. In terms of action orientation, memorial plaques are researched and erected with the young people, and a Jewish cemetery is maintained. However, as the teachers are often restricted by the general conditions (for example, schedules and cross-class groups), which sometimes makes it impossible to plan an excursion, they have found a solution for this as well. They bring the pictures and texts of the memorial site into the classroom, make the tour online, and thus integrate it into the lesson. In this case a follow-up, such as after a real visit, does not take place.

3.5   Follow-Up

All teachers emphasize the importance of follow-up by means such as films, documentaries, and newspapers. They prefer methods such as working in groups and presenting results.[29] In fact, however, for GSP and D follow-up work is minimal. For the project, students were asked to reflect on the visit in an essay, and a group discussion took place. In addition, the memorial visits were no longer discussed in class. This is different with Q, who uses the photos taken by the students to initiate self-reflection and processing of the visit.[30] Based on prepared questions, the young people first deal with the topic by themselves. During this process,

---

[28] Also in this case they have an eye on Protestant church history during the Nazi era.

[29] One teacher (L) puts the emphasis on frontal teaching; otherwise, L believes, the students would not be able to make any connections.

[30] Through the photos, students can recall the place and determine their proximity/distance to it themselves. In many cases, this is what makes reflection possible in the first place.

the young people's own family stories and ambivalences come to light, which are also present in the teacher's mind and are discussed in a serious dialogue. Traditions and images of history are made visible, migration[31] is addressed, and patterns of explanation for backgrounds are sought. In religious education, too, almost all teachers carry out follow-up work in the form of discussions. The dialogues serve the exchange of sensitivities, the passing on of information, the answering of questions, and self-reflection. Where spoken language is not the appropriate medium of expression, pictures are painted, collages are made, or texts are written. This often leads to the opening of new topics, which are dealt with in further texts, pictures, short films, and documents. Possibly this reflects the RE teachers' general process-oriented approach to their teaching. Here too, it is important for RE teachers to underline the connection to Protestant church history. In one case, the debriefing does not take place after a time interval in the next RE lesson, but is held in the coffee house after the visit to the place of remembrance.

4   Summary

Among all interviewed teachers, there is a high level of awareness of educational interventions during visits to memorial sites. All teachers agree that student participation in the excursion must be voluntary. The lack of self-reflexive processes lamented by the historian Pampel (2011) and the self-reflexive factual analyzes demanded by Zülsdorf-Kersting (2011) are tremendously supported by most RE teachers and the art teacher, who generally have very similar didactic considerations. Teachers of GSP and D, on the other hand, are more committed to the competence model, which Meseth (2015) sees as moralizing the topic of National Socialism and which—for him —does not go deep enough. It is difficult to say to what extent these teachers are pressured to implement the model in class. In contrast, a holistic approach is taken in the subjects of religion and art education. Cognitive, emotional, and action-oriented approaches are not only named as important, but are also implemented in combination. Teachers recognize the importance of dealing with one's own family history (Sternfeld 2003), both for themselves and for the students. The interviewed RE teachers especially define their personal access to places of remembrance through their family history. It is all the more astonishing that they do not encourage their students to do so as well, although they strongly ground their teaching in relationship and communication.

---

[31] This topic is exclusively addressed by Q.

Biographical work, on the other hand—when it comes to victims/perpetrators—is the pivotal point of the field trips. This is intended to encourage a change of perspective and empathy. Preparation and follow-up work are recognized as essential elements of memorial site visits. Nevertheless, there is a big gap between theory and practice, and claim and implementation, especially in the subjects GSP and D. In the teachers' opinion, many general conditions, such as lack of time, are arguments against implementation, but still they carry out field trips. All in all, it is clear that the teachers do not indulge in arbitrary didactics and that they think their lessons through well; however, GSP and D teachers only implement their didactic considerations minimally in practice. It is also interesting to note that in a multicultural society, a multicultural approach is not included in the lessons and is hardly mentioned.

These are initial findings about teaching the topic of National Socialism from the teachers' perspective. However, in order to learn more about the teachers' approach and role in teaching Holocaust education, further studies are necessary.

References

Boschki, R. (2001). Zugänge zum Unzugänglichen. Religionspädagogik nach Auschwitz. In O. Fuchs, R. Boschki, & B. Frede-Wenger (eds.), *Zugänge zur Erinnerung. Bedingungen anamnetischer Erfahrung. Studien zur subjektorientierten Erinnerungsarbeit* (346–371). Münster, Hamburg, London: LIT.

Boschki, R. (2015). Erinnerung/Erinnerungslernen. www.bibelwissenschaft.de/wirelex/das-wissenschaftlich-religionspaedagogische-lexikon/lexikon/sachwort/anzeigen/details/erinnerungerinnerungslernen/ch/17f3508aa7e5204858639a9e68cd547a/. Accessed 09 November 2016.

Danner, S. (2020). *Niemals Nummer—Immer Mensch. Erinnerungslernen im Religionsunterricht.* Göttingen: Vandenhoeck & Ruprecht Unipress.

Frölich, A. (2010). Der Diversity-Ansatz als Basis für den Umgang mit heterogenen Gruppen auf Studienfahrten. In T. Hilmar (ed.), *Ort, Subjekt, Verbrechen. Koordinaten historisch-politischer Bildungsarbeit zum Nationalsozialismus* (128–138). Wien: Czernin.

Fuchs, O. (2001). Doppelte Subjektorientierung in der Memoria Passionis. Elemente einer Pastoraltheologie nach Auschwitz. In O. Fuchs, R. Boschki, & B. Frede-Wenger (eds.), *Zugänge zur Erinnerung. Bedingungen anamnetischer Erfahrung. Studien zur subjektorientierten Erinnerungsarbeit* (309–345). Münster, Hamburg, London: LIT.

Hattie, J., & Zierer, K. (2018). *Visible Learning. Auf den Punkt gebracht.* Baltmannsweiler: Schneider.

Herrmann, A. (2007). Erinnerung: Zwischen Vergangenheit und Zukunft liegt die zeitliche Nähe. Anmerkungen zur zeitlichen Multiperspektivität von Erinnerungsarbeit. In J. Birkmeyer, T. Kleinknecht, & U. Reitemeyer (eds.), *Erinnerungsarbeit in Schule und*

Gesellschaft. Ein interdisziplinäres Projekt von Lehrenden und Studierenden der Universität Münster in Zusammenarbeit mit dem Geschichtsort Villa ten Hompel (15–33). Münster: Waxmann.

Kaiser, W., & Rinke, K. (2015). Zum Verhältnis von historischer und politischer Bildung in Gedenkstätten für die Opfer des Nationalsozialismus. In E. Gryglewski, V. Haug, G. Kößler, & C. Schikorra (eds.), *Gedenkstättenpädagogik. Kontext, Theorie und Praxis der Bildungsarbeit zu NS-Verbrechen* (147–165). Berlin: Metropol.

Krondorfer, B. (2013). Interkulturelle Erinnerungsarbeit als offener Prozess. In H.-F. Rathenow, B. Wenzel, & N.H. Weber (eds.), *Handbuch des Nationalsozialismus und Holocaust. Historisch-politisches Lernen in Schule, außerschulischer Bildung und Lehrerbildung* (481–497). Schwalbach/Ts.: Wochenschau.

Lehrke, G. (1988). *Gedenkstätten für die Opfer des Nationalsozialismus—Historisch-politische Bildung an Orten des Widerstands.* Frankfurt a. M.: Campus.

Mayring, P. (2015). Qualitative Inhaltsanalyse. Grundlagen und Techniken. Weinheim, Basel: Beltz.

Meseth, W. (2015). Gedenkstättenpädagogisches Handeln. Zur Etablierung eines Arbeitsfeldes zwischen Professionalisierung und Standardisierung. In E. Gryglewski, V. Haug, G. Kößler, & C. Schikorra (eds.), *Gedenkstättenpädagogik. Kontext, Theorie und Praxis der Bildungsarbeit zu NS-Verbrechen* (98–110). Berlin: Metropol.

Pampel, B. (2011). Was lernen Schülerinnen und Schüler durch Gedenkstättenbesuche? (Teil-) Antworten auf Basis von Besucherforschung. *Gedenkstättenrundbrief*, 162(8), 16–29.

Pollak, A. (2011). Die Verknüpfung von historischem Wissen über Menschenrechte—Herausforderung für Gedenkstätten und Schulen. In B. Pampel (ed.), *Erschrecken—Mitgefühl—Distanz. Empirische Befunde über Schülerinnen und Schüler in Gedenkstätten und zeitgeschichtlichen Ausstellungen* (237–255). Leipzig: Leipziger Universitätsverlag.

Ritscher, W. (2013). *Bildungsarbeit an den Orten nationalsozialistischen Terrors. "Erziehung nach, in und über Auschwitz hinaus."* Weinheim, Basel: Beltz Juventa.

Rosenthal, G. (ed.). (1997). *Der Holocaust im Leben von drei Generationen. Familien von Überlebenden der Shoah und von Nazi-Tätern.* Gießen: Psychosozial-Verlag.

Sauerborn, P., & Brühne T. (2012). *Didaktik des außerschulischen Lernens.* Baltmannsweiler: Schneider.

Schellenberg, M. (2015). Gedenken als pädagogische Aufgabe. In E. Gryglewski, V. Haug, G. Kößler, & C. Schikorra (eds.), *Gedenkstättenpädagogik. Kontext, Theorie und Praxis der Bildungsarbeit zu NS-Verbrechen* (127–146). Berlin: Metropol.

Schwendemann, W. (2013). Erinnern und Lernen in bildungswissenschaftlicher Perspektive—Was soll in einer zeitgemäßen Form der Auseinandersetzung mit dem Holocaust gelernt werden? In W. Nickolai, & W. Schwendemann (eds.), *Gedenkstättenpädagogik und Soziale Arbeit* (101–115). Berlin: LIT.

Sternfeld, N. (2003). Was heißt hier selbstbestimmt? Anspruch und Wirklichkeit von Geschichtsvermittlung in zeithistorischen Ausstellungen. In Büro trafo K.R. Höllwart, C. Martinz-Turek, N. Sternfeld, & A. Pollak (eds.), *In einer Wehrmachtsausstellung. Erfahrungen mit Geschichtsvermittlung* (87–103). Wien: Turia+Kant.

Wagensommer, G. (2009). *How to teach the Holocaust. Didaktische Leitlinien und empirische Forschung zur Religionspädagogik nach Auschwitz.* Frankfurt a. M.: Peter Lang.

Young, J.E. (2004). After Images of the Holocaust in Contemporary Art. In *Ausstellungskatalog, After Images. Kunst als soziales Gedächtnis* (17–24). Bremen: Neues Museum Weserburg.

Zülsdorf-Kersting, M. (2011). Historisches Lernen in der Gedenkstätte. In B. Pampel (ed.), *Erschrecken—Mitgefühl—Distanz. Empirische Befunde über Schülerinnen und Schüler in Gedenkstätten und zeitgeschichtlichen Ausstellungen* (171–192). Leipzig: Leipziger Universitätsverlag.

Links

www.yadvashem.org/de/education/newsletter/10/holograms-and-remembrance.html. Accessed 01 January 2020.

Poland & Latvia

The Auschwitz Concentration Camp

A Journey into Terror

Emilia Smechowski

Abstract

In researching this piece, Emilia Smechowski conducted in depth interviews with twelve of the tour guides at the Auschwitz-Birkenau Museum and Memorial. She wanted to understand the nature of the guides' work, how they handle the sheer masses of visitors, and how they are trained and prepared for this demanding job. Smechowski focuses in particular on Elżbieta Pasternak, a guide who is well regarded by visitors and colleagues for her professionalism, and for her knowledge of the historical details. Smechowski evokes the atmosphere of the place and its history, in addition to giving an impression of daily life for the guides and the current residents of Oświęcim (renamed Auschwitz during the German occupation of Poland). A natural progression of topics related to Auschwitz unfolds in the narrative. Among them are life in Oświęcim today in the face of mass tourism to the camp and the city, how the memorial site is managed, and questions about the preservability of the site and of the objects in its exhibits. These topics are in turn embedded in larger questions about the Polish government's attitudes towards the camp and the history of antisemitism in Poland. The piece reveals much about the current state of Auschwitz, leading the reader to ponder questions about its future as well.

Translation by Julie Winter. The text was first published in a German magazine (ZEIT-MAGAZIN 5/2019: 23. January 2019).

E. Smechowski (✉)
Berlin, Germany
e-mail: Emilia.Smechowski@zeit.de

Seventy-four years after the liberation, the former Auschwitz concentration camp has become a tourist attraction. More than two million visitors now come every year. They are led through the site by guides who are supposed to be morally upright, as pure "as crystals." How do these women and men experience the onslaught of the masses? Emilia Smechowski met with them at their work over a period of several months.

Most cry when they see the hair. Elżbieta Pasternak remembers the shock it gave her when she first saw it at age thirteen. She never wanted to come back to this place. Today she works here. Making people cry is not her goal, she says. She has to take care of the tour. Is everyone present? Can everyone hear?

Elżbieta Pasternak, forty-five years old, known as Ela, has packed her bag: cough drops, water, a flashlight, her cell phone. And an umbrella, but not for pointing the way—he actually needs it for the weather. She's had coffee and eaten two sandwiches. She rarely gets a lunch break, she says. A good breakfast is, therefore, the most important thing before a shift.

Line 1 takes her from home to the main camp. The parking lot is already full of buses. As one of 328 guides, her job is to explain Auschwitz. "Guide," not "Führer," is used here in German.

Elżbieta Pasternak reports to the reception. Her group has not yet arrived. She sits on a bench in the entrance hall and pulls her long wool coat tight. Visitors surround her, Italians wearing soccer jerseys, a tour group from the Philippines. People come to Auschwitz from all over the world. The number of visitors has almost doubled in the past ten years. In 2018 it was 2.1 million.

You meet people here who had never heard of Auschwitz until the day before they came. Like the Melbourne tourist group that has just arrived in Krakow, about sixty kilometers from Auschwitz. They don't know that a right-wing conservative party rules Poland. They're here because their hotel offered them the chance to spend a day in Auschwitz-Birkenau for the equivalent of thirty-six euros, traveling "in a comfortable Mercedes-Benz van, with Wi-Fi and air conditioning," as the brochure stated.

Other groups, especially school classes, are better prepared. They've given presentations on the subject and have read Anne Frank's The Diary of a Young Girl. And then there are German visitors like the ones Elżbieta Pasternak guides. She's happy to take the German-speaking tours, she says. "I don't have to explain the Holocaust to the Germans."

Almost all visitors come here only once in their lives. Almost all visitors book the shortest tour—three and a half hours. What they see and hear in Auschwitz during this time is what will stay with them.

Elżbieta Pasternak's group is here—twenty-nine people from all over Germany, everyone from students to retirees. Before entering the site, they have to attach a yellow sticker to their chest. Everyone is tagged so that the groups stay together. Green is for French, orange for English. In Auschwitz you can book guided tours in twenty languages. There are no self-guided audio tours. "Can everyone hear me?" asks Elżbieta Pasternak. She speaks into a small microphone that hangs around her neck. "There's a little wheel on the side for adjusting the volume." Ever since the visitors can hear guides' voices through headphones, it's said to have become somewhat quieter in Auschwitz, and no one has to scream anymore to be heard.

As she heads toward the famous gate with her group, Elżbieta Pasternak asks that visitors not smoke, eat, or make phone calls on the premises. At this moment, almost everyone pulls out their cell phone and takes a picture of themselves in front of the lettering "Arbeit macht frei." "We are now on the site of the former Auschwitz I main camp, which was founded in 1940," says Pasternak—and with that their tour officially begins.

Elżbieta Pasternak was born in Oświęcim, the name of the Polish city where the camp was located before the war. It still goes by that name. Pasternak studied German language and literature in Kraków. She loves Thomas Bernhard. German literature, she says, is dark but deep. This is the spirit she feels when she reads, the deep spirit of the past.

She explains that the city was chosen as a location for a concentration camp by the Nazis because of the good rail connection and because there was already a Polish barracks here. "Suddenly Oświęcim was called Auschwitz, the market place was named after Adolf Hitler, and the people who lived here were largely resettled. The Jews ended up in ghettos. Many came back later—to the concentration camp." Elżbieta Pasternak doesn't like to talk when walking. She wants to look people in the eye.

Ela, colleagues say, doesn't just rattle off her knowledge; she stays engaged. There are teachers who come with their classes every year and always try to book "the Pasternak woman," even though they know that this isn't possible because guides in Auschwitz are randomly assigned. Actually, they are all supposed to tell the same story—but if you go to Auschwitz several times within a year, sometimes officially as a journalist, sometimes unofficially as a tourist, you will find that some guides seem bored as if they were doing a school presentation. Others tell a lively story, and still others try to shorten the tour to get home faster.

"I couldn't do it every day," says Elżbieta Pasternak. She guides tours through Auschwitz about ten days a month. In addition, she works at the International Youth Meeting Center, where she discusses guilt and responsibility with young people from all over the world. This change in routine, she says, prevents her from getting into a rut. "Auschwitz is like standing before a class of students. It takes a lot of energy to reach a group of thirty people."

Visitors are lining up outside Block 4. Block 4 is always jammed. That's why most guides don't like it. Elżbieta Pasternak thinks that Block 11 is even busier—that's where the death cells were. She is now coordinating with a colleague about who can be the first to go in. The colleague taps on her watch in slight despair—no problem, Pasternak waves her and her group through.

In the first room of Block 4 there is a map of Europe with the places from which people were brought to Auschwitz. "They had no idea what to expect," says Pasternak. "In contrast to the Polish Jews who had already gone through various ghettos, those from the West really believed they were being resettled." Most of them were clueless, bringing whatever they could fit into their suitcases. Loot for the Nazis to steal.

"We're going upstairs now. Please walk up the stairs in single file, on the right side, stay all together," says Pasternak. Up above, behind a glass wall there's a model of a gas chamber and a crematorium. Behind another, empty Zyklon B cans. Elżbieta Pasternak is no longer frightened, she has seen it all far too often. "Zyklon B, you may know, was produced by the German Pest Control Society." In Auschwitz, more Zyklon B was needed for delousing clothes than for gassing people. A man asks how many died from a "portion." "How many people?" Pasternak asks more precisely. "Up to 2,000. And now, in this next room, we ask you not to take photos out of respect for the dead."

Then they stand in front of the sea of hair cut from the prisoners' heads to be sold to German textile and carpet factories at 50 pfennigs a kilo. Over the decades, the hair has turned a muddy gray—it looks as if it could crumble to dust at any moment. Everyone is silent; a man in motorcycle gear is standing still. He cannot break free. Elżbieta Pasternak remembers the first time she was here herself with her school class. The hair scared her. She recalls, "This fear somehow went right to my gut." Today she is no longer moved, but she feels empathy towards her visitors. She gives them some time with the hair.

In Block 5 there are personal belongings of the prisoners. Here she says little. Don't the piled-up suitcases and the children's dresses explain themselves? The hair and toothbrushes, the enamel dishes, glasses, mountains of shoes? "Look, platform ankle boots," says one friend to the other. "Who would have thought they were wearing those back then."

The past and the present constantly collide in Auschwitz. If you can't bear the baby dresses, in the next moment you might notice the sweet smile of a prisoner in one of the photos. You might feel like throwing up, and a few minutes later your stomach growls. "I understand that," says Elżbieta Pasternak. "People are still people, even here in Auschwitz."

The preservation department ensures that the objects in Block 5 look as if they had just been left behind by their owners. There's no object in Auschwitz, not even a beam from a wooden barrack, that has not been not worked with in some way. The preservationists take each shoe individually in their hands, clean it carefully, and grease it.

Aesthetics do not count in Auschwitz, the curators say. Nothing should be made beautiful here. What matters is authenticity—and the question of at which point it is lost. If a suitcase has a crack, should it be repaired? No, because a crack can mean that the owner could have been frantically searching. Crusted dirt on the shoe? It stays because maybe it was raining and the person was standing in the mud. "We are touching the trace of a human being," says the head of the department. "It's almost as if we're touching the person through it. That's why we put out as many shoes as possible. Who decides whose story is more important?"

The preservationists can hardly keep up with the work. When the suitcases are done, a barrack is in danger of collapsing. Then they have to completely dismantle and rebuild it so that it looks the same as before. They start every morning at seven, hunched over paper, stone, metal, wood, and leather.

Everything can be preserved. Only the hair of Auschwitz will not last much longer. Experts from all over the world don't know the solution. Eight more years, say the guides, and then the dead people's hair will no longer exist. The museum doesn't want to confirm this figure, but also has no other estimate—the subject is a "delicate matter," says a staff member on the phone.

Preserving, not tearing down, was the wish of the former prisoners after the camp was liberated on January 27, 1945. Everything should remain as it was. Two years later, some of them opened the memorial and initially led visitors through the barracks themselves. The state of Israel was not yet established, and Germany and the Allies were busy rebuilding and prosecuting the perpetrators. Poland alone took care of preserving the camp.

If you ask historians and archaeologists about the future of Auschwitz, they all agree: At some point, Auschwitz will have to change. Authenticity weakens the more a shoe is greased, the more the hair disappears. So instead of just archiving the past, they say we should also talk about the present in Auschwitz. However, they also maintain that as long as there are survivors, the memorial will remain as it is.

American Judaist James E. Young, who has been researching the topic of memory in memorial sites for over thirty years, says, "Auschwitz will change. We also need to talk about the environment that made the Holocaust possible.

About the normal civilian population, about the Allies looking the other way for far too long. This is the only way we can bring the topic into the present. There is only one dilemma if we allow more gray areas—are we opening the door for Holocaust deniers?"

Because Auschwitz is more than just a symbol. It's still a crime scene. In the trial of David Irving, a Holocaust denier in the late 1990s, the ruins of the crematoria served as evidence: There were the gas chambers. There were over a million dead.

Furthermore, the tourists want it, they want the "real" concentration camp. Auschwitz has a record-breaking number of visitors, and it's also considered a place of "dark tourism" where you can go to be scared. Like at Ground Zero or in Chernobyl. Some visitors want to go into the gas chamber right at the beginning, others ask what it is like to suffocate.

A few weeks ago Pasternak had to call security. "You stupid Nazi!" a man had shouted because a woman had asked him to move out of the way so she could see a shoe. "It was the first time I ever heard anyone in Auschwitz shout in German. I wouldn't have thought that was even possible," says Pasternak. "That really affected me." The man later apologized.

Mostly she finds her work physically, rather than mentally, demanding. "We rush across the grounds, it's always too crowded, rarely can you stop to explain something a little longer. In the evening, when I've had two groups in a row, all I do is cook and put my feet up. And I'm starting to wonder: Does the quality suffer due to the masses that move through?"

Large photos in Block 6 show what life in Auschwitz did to the prisoners' bodies. Taken by the Russians after the liberation, they show young people, ancient bodies, thirty-one years old, twenty-five kilos, skin and bones. Some in the group have to leave the room immediately. It is a fine line, and every group is different. Auschwitz is based on emotions, on the power of such images. Some are moved, others are paralyzed, some giggle, others bend over their cell phones. Add to that the constant waiting, the large number of people and, yes, also the evil and sadness in the display cases—all of this is very draining. Many visitors realize too late that they did not have enough breakfast. They now have to hold out until the end, and of course they do hold out. How could they dare bite into their candy bar here of all places? They are only allowed to eat outside the camp, in the canteen or in the snack bar next to the exchange office.

Elżbieta Pasternak is already standing at the iron door of Block 11, when another group suddenly pushes forward. Pasternak says to her colleague in Polish, "I'll just go in, do you mind?" She would never show more annoyance than that at Auschwitz—now she quickly disappears into Block 11 with her group behind her. Her colleague tries not to look angry.

Not only were there detention cells in Block 11, the first gassing attempts also took place here. Reception in the basement is poor. Over the headphones you only hear fragments of Elżbieta Pasternak's narration: "one by one [...] very narrow [...] standing bunker [...] Father Kolbe [...] sacrificed himself [...] starved." There are candles in the cell, and rosaries are draped around them.

The Catholic priest Maximilian Kolbe went to the infamous "hunger bunker" in place of another Catholic prisoner in 1941. When he was still alive after two weeks, the Nazis murdered him with a phenol injection. He was canonized forty years later. A hero, a martyr, that's one part of the story. The one that the guides tell at Auschwitz.

Each guide must first pass a written, then an oral, and finally a practical exam. Written instructions indicate what is to be covered where. In front of the map of Europe: where the prisoners came from. At the hair: the practical uses. In the gas chamber: the killing procedure. The guides should not only convey facts, but also the stories of individual prisoners. The management of the museum decides which stories should be told, not the guides. It specifies the route and the corresponding narrative.

The fact that Maximilian Kolbe published high-circulation magazines full of antisemitic texts before the war is not mentioned in Auschwitz, for example. "World Jewry," it says in one, gorges itself "like a cancer on the national body," of the Poles, of the country where about half of European Jews lived in 1939.

Political psychologist Michał Bilewicz, who researches stereotypes and dehumanization of minorities in Warsaw, says that Kolbe shows how words become deeds. "The stereotype of a Jew, to which his magazines also contributed, demonstrably prepared the ground for pogroms. Like in 1941, when Poles murdered between 300 and 400 Jews in Jedwabne."

Those who want to shed light on the dark spots in Polish history have chosen bad times. Ever since Poland has been governed by a right-wing conservative party, pressure on museums, newspapers and schools is growing. Poles have always found it difficult not to view themselves simply as victims of National Socialism or as heroes of the resistance. The right-wing conservative Law and Justice party (PiS) government even enacted a law a year ago that made it a punishable offence to hold Poles jointly responsible for the Holocaust. The world was outraged, and the law has since been changed.

The seeds have long since sprouted. In February 2018, a regional education politician argued on an online right-wing platform that only Polish guides should conduct tours through Auschwitz. The memorial currently employs 320 Poles and eight foreigners, and there have never been many more than that. One of them, an Italian, then became a victim of right-wing agitation. "Poland is for Poles" was spray painted a few weeks later on the entrance to his Krakow apartment, and "No foreign guides in Auschwitz!" Also a star of David and a swastika.

Auschwitz Museum is under the supervison of the Polish Ministry of Culture. The director and his deputies are appointed by the government. Since 2015, when the new government came into power, the tour of Auschwitz has not changed. However, starting in 2020, a new exhibition is planned, which they say will be constructed in stages. And the current director's contract expires at the end of 2019.

Elżbieta Pasternak and her group have left the basement; there is time for a short commemoration at the death wall outside. This is where prisoners were shot—today politicians place wreaths of flowers here. Last summer the musicians Kollegah and Farid Bang also stood here, guiltily bowing their heads after they had rapped antisemitic lyrics. Half of Germany again wondered how antisemitic German hip-hop and German society were, and whether it wouldn't be a good idea to make a concentration camp visit compulsory for school classes.

School classes from Germany usually stay overnight. They are accompanied by teachers, some of whom have been coming for twenty years and who firmly believe that visiting a concentration camp can make a difference at a time when they are observing an increase in swastika painting in classrooms. And it looks as though they're right. In the evening, at the reflection session, everyone says, "It's good we were here. Our friends should go too." They may not remember the historical dates, but they do bring away a feeling—one of responsibility.

Elżbieta Pasternak believes that nobody should be forced to go to Auschwitz. The place is too hard for that. She is now standing next to a gallows. This is where Rudolf Höß, the commander of Auschwitz, was hanged in 1947. It was determined that he should die in the place where he had people gassed and where, a few meters away, he had dinner with his family after work. In the summer he sometimes splashed around with his children in the river Soła—if it wasn't black from human ashes.

You can see the old Höß house from the camp; it's gray and has two-stories and a corner balcony. It is not part of the memorial, and most guides are very annoyed whenever the groups want to know who is living there. "A Polish family," they simply say, ignoring the horrified look on the visitors' faces that asks, "But how can they?"

If you leave the camp and ring the doorbell to this house, a young man opens the door. He is thirty-six years old and works in a pet shop; his wife is an English teacher. He quickly closes the door. He no longer wants to talk to journalists who just want to make a creepy story out of his home. He sees nothing wrong with living here.

Like most people here, he wants to live in peace, in the present. And the present is not called Auschwitz, it is called Oświęcim. This difference is important to the residents. After all, their small town is like many others. There are advertisements everywhere for loans and courses in Capoeira, in summer people sit in ice cream parlors, on Halloween ghostly skeletons roam the city. There are two discos, but most young people prefer to drink in parking lots or to party at McDonald's with strawberry shakes.

Anyone who visits Auschwitz and not Oświęcim will not notice all this—the camp is on the outskirts of the city. Those who visit Auschwitz usually leave immediately. There is a hotel directly opposite the museum called the Imperiale. The young women who work at the Imperiale say that here they can quickly see what a person is made of. The bitter ones are even more bitter, the kind ones even kinder. There are people who tip a lot and people who ask for a room facing the street so that they don't have to look at the old tracks. Some angrily call the front desk when they hear young people roaming the streets at night. How could you listen to techno so close to a concentration camp!

When everyday life of Oświęcim reaches the outside, the world is happy to be outraged. Then newspapers around the world pose questions about morality. A disco "in Auschwitz" or a McDonald's? As if people were dancing or gobbling down Big Macs in the barracks. This annoys Elżbieta Pasternak too. She doesn't want to live anywhere else, she says. Oświęcim is her city. Her family lived here before the war. What do they remember about the camp next door? "Unfortunately, I don't know exactly," she says. "I never asked my grandparents. My grandmother once talked about the train cars that they could hear screams from and the fact that the cars came back empty. She had Jewish friends who stayed away from school from one day to the next. I didn't look into it further. I regret that. It's hard to talk about such things in the family."

Before the war, Oświęcim had 14,000 inhabitants, including around 8,000 Jews. Today there are almost 40,000, no Jews. Foreign visitors pray in the synagogue.

It has started to rain; the paths turn to mud, and Elżbieta Pasternak's group swiftly heads for the gas chamber. Inside, everyone first has to get used to the dark. In the gas chamber, Elżbieta Pasternak is silent. She points her hand towards the ceiling, there, there and there. Like a flight attendant who points to the emergency exit, she shows the vents through which the Zyklon B trickled.

She had already described outside what happened after the doors were locked. "When the prisoners realized something was taking their breath away, they started to climb on top of each other in panic. The poison rose from floor to ceiling, and so the strong ones landed on the top, the weak, especially children and old people, on the bottom. About twenty minutes later the screams, the pounding on the door, stopped. The Sonderkommando opened the door and began removing and burning the bodies."

The ceiling is black and peeling. Suddenly it is so quiet that you can hear the raindrops falling. And when one of the group is hit by a drop, he gasps out of fright. A few start giggling, Elżbieta Pasternak turns around for a moment, and then there's silence. She directs the group to the next room, to the ovens.

The vast majority of the 328 guides live in Oświęcim and in the surrounding villages. They are forty years old on average, just over half of them are women. There are historians among them and primary school teachers and those who found jobs right after finishing school. They are paid per tour; for a tour of three and a half hours they get the equivalent of around fifty-six euros, which is good money in Poland. The memorial has become the largest employer in the region, next to the chemical plant.

Under communism, Auschwitz was as closed as the country in which it was located. Tourists rarely came; if they did, then it was mostly from the German Democratic Republic or the Soviet Union. Scientists and archaeologists who were there at the time tell of the icy tone that prevailed. The then director of the memorial, a former political prisoner, was jokingly called the "Commander of Auschwitz." What was told came from the very top. The history of the People's Republic of Poland went as follows: "Catholic Poles and people of other nations died here." The Jews as a group of victims virtually did not exist, although they made up ninety percent of the victims. The number of dead given—four million— was also wrong. When historians from all over the world met in Oświęcim and Oxford after the fall of the Iron Curtain to redesign the direction of the exhibition, a bitter struggle began between Poland and Israel. Are the Poles allowed to put up a cross on the site? Are the Israelis allowed to come with their flags and pray? A competition arose among victims, which revolved around the question: Who owns Auschwitz?

Even today, Auschwitz, the museum, is not just a place of silent remembrance. Anyone who moves alone on the site is eyed suspiciously by the museum staff, as is anyone who stands still somewhere for a longer period of time. Except for Elżbieta Pasternak, no guide initially wants to talk about their work. The museum administration had promised to find people to interview, but in the course of the year, guides say that they never received the request from ZEITmagazin. Only gradually have they begun to speak out; most of them ask not to be mentioned by name. They're afraid.

Years ago, a female guide wrote on a private blog how stressful the work in Auschwitz could be. She didn't get any more assignments after that. The museum says the decision had nothing to do with the blog entry, and it would "not comment further on personnel decisions." Some guides say that an employee of the press office regularly checks what the guides post privately on Facebook and Twitter —the employee is called "the spy" on the inside. "We don't monitor private accounts," replies the museum. "But we do point out to our guides that private postings could be interpreted as statements by the museum."

Historical questions are readily answered in Auschwitz, current ones less so. What influence does the government have on Auschwitz, and to what extent is the shift to the right in society noticeable here too? When the attack on the Italian guide became public, the management issued a statement in support of him. There was support from the colleagues—but he reports that they also circulated a rumor that he may have spray painted his own door in order to sell his books about historical Krakow better. He was very disappointed about that, he says.

The administration of the museum is in a former SS barrack. The Nazi pharmacy used to be located here, the dentist, the German canteen—a small figure at the entrance sitting on a beer barrel testifies to this. Otherwise gray plastered walls and concrete; in one of the rooms there is a huge old aerial view of the site. It looks as if time has stopped.

In the director's office the windows are taped over; no one is supposed to look in from the outside. Piotr Cywiński, forty-six, is rarely on site. He lives with his family in Warsaw and attends conferences all over the world. You have to wait months to talk to him. Now he sinks into a beige sofa and sucks on the straw of his cup of mate. He studied history, focusing on the Middle Ages, and speaks six languages. He wears jackets only on official occasions, today he's wearing a Harley-Davidson shirt and a hooded jacket.

When Cywiński became director in 2006—the PiS party was already in power then—Auschwitz was run down and almost broke. There was no money for a complete restoration. "So I decided to invest in conservation," he says. "Otherwise we would have had nothing left to exhibit. Everything else could wait." The museum is now in a good financial position. This is also due to a foundation, founded in 2009 on Cywiński's initiative, to which various countries contribute. Its endowment is currently 120 million euros.

Cywiński belongs to the Catholic elite in Poland. He is said to have good contacts with the ruling party. It is difficult to classify him politically. "I have already experienced several ministers of culture and have had no problem with any of them." He won't say more than that. He does not yet know whether his contract will be extended at the end of 2019. He encourages his staff not to comment on political issues under any circumstances. "For me, Auschwitz is a historical, not a political place," he repeats again and again.

Yet in times of growing nationalism and antisemitism Auschwitz is more political than ever. In March 2018, Piotr Rybak, probably Poland's most famous racist and antisemite, visited the memorial. He had just been released from prison after burning a doll that was supposed to represent an Orthodox Jew in a market square in Wrocław. Now he walked through Auschwitz with his retinue and spoke his message to the camera: Over 4.5 million Poles had died in this place; everything else was Jewish lies. The museum knew about the visit.

When Alternative for Germany party politician Björn Höcke called for a "180-degree turn" in the politics of remembrance and then wanted to visit the concentration camp memorial in Buchenwald, he was banned from going there. The management in Auschwitz didn't dare to do this with Rybak. Or didn't want to. According to the museum, everyone is allowed to go to Auschwitz, but insignia such as swastikas are prohibited. Members of the Polish far right, however, rarely appear with a swastika.

The shuttle bus from Auschwitz to Birkenau, the second camp near the main camp, departs. The bus is yellow, a donation from the Berlin public transportation company. Elżbieta Pasternak holds on to a loop from the ceiling of the bus. There is a German word on the floor below her: "Loading ramp."

The man in motorcycle gear has been trying to ask her a question the whole time, but he can't get through—it's too crowded. He wants to know exactly what the Allies knew about Auschwitz and the gas chambers, whether the worst could not have been prevented. He postpones the question until later.

"I feel bad that there is often so little time for dialogue," says Elżbieta Pasternak. After a guided tour, she usually has to hurry back to the entrance, where her second group is already waiting. "I don't know how to change the situation. One would have to drastically reduce the number of visitors. But isn't it also important to get as many as possible to see Auschwitz?"

Elżbieta Pasternak works almost every day, including weekends. She lives alone in a two-room apartment in a prefab building. Sometimes she rides a bicycle along the river with her niece. After that, she often stays with her sister for dinner. They don't talk about Ela's work. The sister works in a company that installs windows in houses. "What's she going to ask me? Ela, how was your day in Auschwitz?"

After three kilometers, the bus releases the people into a postcard motif, the entrance gate of Birkenau, through which the trains used to pass. When Elżbieta Pasternak leads the visitors up to the watchtower, her group cannot believe it:

It's so big, 176 hectares—the vastness here frightens everyone. Even from above there is no end in sight.

Just a few years ago, there were no guards at Birkenau. The village youth played ice hockey on the frozen pools in winter. People took Zyklon B cans to store motor oil; some knocked boards out of the barracks. Today the site is tidied up, a security guard drives past on a Segway, and it's now out of the question for young people to play ice hockey there.

The writer Ruth Klüger would have liked these kids. As a child, she was interned here in the so-called "family camp." She never wanted to go back to the memorial, curated, all made up. "What would I do there?"she writes in her book *weiter leben.* "Whatever you think you'll find there is probably something you already brought with you."

Suddenly, in the vastness of Birkenau, a male voice begins to sing a prayer. A calm, sad melody. Over and again the singer pauses, then starts again. "The Kaddish?" whispers one of the group; Elżbieta Pasternak nods. A young rabbi is singing the Jewish prayer. He has placed a small amplifier where the forest begins and the memorial stands.

Elżbieta Pasternak says she thinks it's good that Jews pray here, but some guides roll their eyes when they see the ceremonies. Some of them complain over a coffee in the canteen about "the Jews" who "make themselves important," without noticing that a reporter can hear what they are saying. Two days later a female guide stands in front of the office, peels an orange and says to her colleague: "I have Jews again in a few minutes, unfortunately."

Everyone who works in tourism is sometimes annoyed by tourists. Still, if you spend a lot of time in Auschwitz, you won't hear similar talk about any other group. A few guides confirm that there is always resentment about "the Jews." Elżbieta Pasternak says that she hasn't noticed it so far. "But I know that we guides are expected to be morally pure, like crystals. That's not so easy. Why should people be better here than anywhere else?"

Did people think that Auschwitz was an island, untouched by the outside world? If Poland, if the world, has a problem with antisemitism and xenophobia—then Auschwitz does too. The museum willingly answers my final questions. However, the museum leaves out whether the management has heard talk like this and how they deal with it. They don't respond to any further questions.

Elżbieta Pasternak has arrived at the Birkenau exit. "We'll stop here now," she says. "Thank you for your time." Her tour is over, all the sentences have been spoken. A bus will take the group back to Krakow in a few minutes. And when the Germans clap shyly, she puts her right hand on her chest and smiles.

The survivors did everything they could to hold onto their memories for posterity. In Yad Vashem alone, the memorial in Jerusalem, there are 51,000 videos of eyewitness talks. Once you start dealing with the topic, it is difficult to stop. Elżbieta Pasternak knows this fascination. "We really want to understand the Holocaust," she says, "but we can't. We'll never get there. Closing this gap is what drives me too. That's what keeps us going, isn't it? That we are left with this question."

Emilia Smechowski has visited Auschwitz and Oświęcim seven times. She spoke to twelve guides and took tours in German, French, Italian, English and Polish—her mother tongue. Before embarking on this project, she had been in a concentration camp once, in Sachsenhausen, as a schoolgirl in Berlin.

Guiding at the Jewish Holocaust Sites in Riga

Difficult History, Tourism, and Individual Experiences

Inese Runce and Aija Van der Steina

Abstract

This article analyzes processes of formation and transformation of Riga Jewish heritage and Holocaust memorial sites after the Second World War and the historical transformation of the guides' professional work and education in modern Latvia. The conclusions are based on theoretical studies and Riga guides qualitative research results. Jewish community history and the Holocaust play an important part in Riga's heritage tourism offerings. One of the most important tools for professional Holocaust and Jewish history guides' work is knowledge of both intercultural communication and general knowledge of Jews and Jewish history, which has not been fully covered in the process of guide education and professional development in Latvia.

1   Introduction

The story of Holocaust history and memorials, which has been consistently shared by guides working in Latvia for decades, is not easy to understand without a tour in twentieth century history. The human challenges faced by experts in this field,

I. Runce
Institute of Philosophy and Sociology, University of Latvia, Riga, Latvia

A. Van der Steina
Institute of Philosophy and Sociology, University of Latvia, Riga, Latvia
e-mail: inese.runce@lu.lv

A. Van der Steina
e-mail: aija.vdsteina@lu.lv

© The Author(s), under exclusive license to Springer Fachmedien Wiesbaden   127
Gmbh, part of Springer Nature 2022
A. Ballis (ed.), *Tour Guides at Memorial Sites and Holocaust Museums*,
Holocaust Education – Historisches Lernen – Menschenrechtsbildung,
https://doi.org/10.1007/978-3-658-35818-1_8

the evolution of their professional education and development were also closely intertwined with the history, tragedy, and rebirth of the Jewish community in Latvia. Further, the most extensive professional challenges of guides working in the field of Holocaust and Jewish cultural heritage are closely related to both the tragically rapid forced loss of Latvia's historical memory and its slow return in society over the last 30 years. These professionals encounter the pace of general economic development: ups and downs, existing cultural and economic policies, including tourism, state and municipal policies, existing understanding, and lack of vision about the use of difficult cultural heritage in the tourism business. Problems exist concerning the sustainability of supply and demand of dominant tourism and the fact that the modern world has lost "the ability to even think about public policy outside the confines of a narrowly interpreted economic framework" (Džads 2012). The professional framework of a guide has always been strictly limited on several levels, as a "phenomenon of guiding [that] has been connected to a variety of questions through its intimate relations to innovation, politics and economy, social interactions and power relations, culture and indigenous tourism and local versus global" (Weiler and Black 2015, p. 13).

Until now, issues concerning Jewish history, Holocaust memorial research and communication, memory, and general research of tourism processes in Latvia have been sufficiently highlighted in the academic environment. However surprising it may seem, no research has been carried out in the professional field of guides and their challenges, education, and wellbeing. This article will reveal the experience of guides in historical and modern cross-sections, exploring and interpreting Jewish heritage and Holocaust sites for tourists in Riga.

From an internationally unknown travel destination in the mid-1990s, Riga has become a popular travel destination among foreign tourists. Since the beginning of its independence, Riga's tourism offer has been based on 800 years of history and rich heritage (Druva-Druvaskalne et al. 2006). The most popular attractions of Riga, the Medieval Old Town and Art Nouveau district with attractive architecture, museums, and events (Rozite and Klepers 2012), have become the main elements of place promotion with the aim of creating a place identity that would confirm its shared heritage with Western Europe (Light et al. 2020).

The Soviet heritage, which has attracted the attention of foreign tourists since the fall of the Iron Curtain, draws tourists as well. Nevertheless, it is "little valued by the local population" (Light et al. 2020, p. 469) and mainly includes examples of Soviet architecture such as monumental public buildings and housing, and has only emerged in the city's tourism offerings over the last five years.

The KGB Museum and Occupation Museum, and sites related to the heritage of the local Jewish community and the Holocaust Museum "Jews in Latvia,"

Museum of the Riga Ghetto and Holocaust in Latvia, as well as the memorials Riga Choral Synagogue, Jewish Memorial at Rumbula, Jewish Memorial at Biķernieki, and Žanis Lipke Memorial, reveal painful, unwanted, and complicated historical events in the context of WWII and the Holocaust. They have become an essential component of Riga tourism both in terms of demand and supply. For example, the travel site *Lonely Planet* (2021) ranks the Riga Ghetto and Latvian Holocaust Museum among the top ten choice attractions. However, the sites mentioned above are a small part of the approximately 260 identified sites in Latvia related to Jewish heritage and the Holocaust (Die Jüdische Gemeinde "Schamir" 2015).

Currently, the municipality of Riga offers the services of 447 certified guides (The Investment and Tourism Development Agency of Riga 2021a). Guide certification, setting requirements for professional qualifications and knowledge of foreign languages, has only been in place since 2014. Its goal is to ensure the professionalism of guides and the quality of services provided (The Investment and Tourism Development Agency of Riga 2021b). Guides from the European Union and the European Economic Area are subject to the principle of Free Movement of Services. They must register the provision of temporary services with the Riga Investment and Tourism Development Agency.

2 Jewish Heritage, Holocaust Sites, and Guiding in Riga—An Historical Overview

The Jewish community is historically one of the oldest existing ethnic religious communities in Latvia. The first Jewish community in the then Livonian territory was established at the end of the sixteenth century. However, historically, information about the presence of Jews in Riga has been preserved since the Middle Ages. Jews in Riga and the territory of modern Latvia have experienced the rule of various powers (German, Polish, Swedish, Russian) and various laws implemented in relation to Jews. Most of the modern territory of Latvia (except Latgale) had not been in the settlement zone of the Russian Empire. Historically, two Jewish linguistic communities lived in Latvia: the German-speaking Jews of Kurland and the Yiddish branch of Polish Jewish culture.

After the end of the First World War on November 18, 1918, the Republic of Latvia was proclaimed. Already on November 17, 1918, the Latvian People's Council (LTP) recognized the rights of ethnic minorities, giving them the right to delegate their representatives to legislative and executive institutions and guaranteeing the cultural rights of communities and the preservation of their identity. However, the situation got worse after the establishment of the authoritarian regime of Kārlis Ulmanis on May 15, 1935, when the democratic system was destroyed.

The education laws issued on December 8, 1919 stipulated that minorities have the right to organize and manage their community schools, i.e. exercise their educational autonomy, which created the basis for the formation of the Latvian Jewish minority and its cultural autonomy. This promoted the emergence of Jewish loyalty to independent Latvia in the 20s and 30s of the twentieth century (Dribins et al. 2001). The last census in 1935, which took place shortly before the loss of statehood and the impending Soviet and Nazi occupation, recorded 93,479 Jews living in Latvia. The Latvian Jewish community, with its extensive network of schools and community organizations, had branched out throughout Latvia. Its presence was evident and noticeable in both economic and cultural life. The largest percentage of Latvia's Jewish population had lived in Riga for centuries, which was a large cosmopolitan cultural and economic center in Northern Europe. In 1920, Riga became the capital of the independent Republic of Latvia. In 1920, there were 24,725 people in the Jewish community. During the 1920s and 1930s, the Jewish community flourished: a community council was formed, hospitals and charities were opened, theaters were opened, and newspapers and magazines were published in Yiddish, Hebrew, German, and Russian. During this time, the city had 12 elementary schools, two Jewish gymnasiums, a Jewish high school, and a conservatory. In the 1920s, there were five synagogues in Riga and more than 20 houses of prayer. In the 1920s and 1930s, Riga also became the center of the Zionist movement in Eastern Europe. In 1935, 43,672 Jews or 11.3% of the total population, lived in Riga (Museum "Jews in Latvia" 2021a). The activities of the Jewish community were partially suspended during the first year of Soviet occupation during the period from June 1940 till June 1941; during the Nazi occupation between 1941 and 1945, the community was completely destroyed.

The Jews who survived the Holocaust after the Second World War realized that their tragedy was not given due attention in Soviet Latvia; for example, the authorities did not participate in cleaning up the graves and creating memorials. According to the survivors, the Jewish tragedy was not adequately portrayed in the press, literature, and art, so their feelings for the future were marked by a mixture of relief and sadness, as well as hope and suspicion. The policy implemented by the USSR commemorating the Holocaust balanced between antisemitic Holocaust denial and the accusation of all Eastern Europeans except Russians of cooperation with the Nazis (Bērziņš 2015).

The antisemitism of the Soviet era was compounded by a policy of silence regarding the Holocaust and its remembrance beginning in the late 1940s, in the form of refusals and avoidance of erecting monuments and memorial sites, as well as instructions not to highlight Jewish victims when speaking and writing about the crimes of the Nazi regime. Both attempts to study the course of the Holocaust during the Nazi occupation and attempts to commemorate the victims of the Holocaust, which were classified as attempts to highlight one part of the victims while pushing others aside, and thus condemnable, went beyond any permissible discourse (Bērziņš 2015).

During the Nazi occupation, local authorities were given the task of destroying Jews, Holocaust victims, and graves, thus concealing evidence of the Holocaust. Although these actions were not an expression of deliberate denial of the Holocaust, it is paradoxical that the Soviet government essentially continued this policy by turning old Jewish cemeteries into parks and greenery after the war (Bērziņš 2015). The old Jewish cemetery in Riga, which dates to 1725 and was part of the Riga ghetto during the Nazi occupation, was completely levelled to the ground. Beginning in the 1960s, Riga became one of the centers of the Jewish Zionist movement in the USSR, attracting a large number of Soviet Jews who for this reason moved to Riga, but this part of the Jewish population had a different historical memory and connection to Latvia's Jewish cultural heritage and memorial sites, and the tragedy experienced was indirect.

In 1988, the Latvian Jewish Cultural Society, which symbolically marked the beginning of the revival of the Latvian Jewish community, was established in Riga. In 1989, the first Jewish school in the Soviet Union was opened in Riga, and a historical discussion and research on the history of the Jewish community, the Holocaust, and the identification, promotion, and documentation of this difficult historical heritage was launched. It started with a new wave in the 1990s of the twentieth century after the collapse of the USSR and the restoration of Latvia's independence.

In Riga, the Jewish community informally resumed its activities after the Second World War. There were two synagogues, but the main gathering place, no matter how tragic it sounded, was the cemetery and the place of killing Jews in Riga in Rumbula, but later also in Biķernieki. The first informal guides for commemorating events and visiting the Holocaust were mostly Holocaust survivors from Riga or their relatives, in a way combined two responsibilities: amateur research and storytelling functions. As Bērziņš emphasizes, the Soviet authorities were informed about these activities; however, the then ideological concept of the USSR did not provide for the mention of victims, because the discourse of victory dominated politically, which did not include the dimension of victims (Bērziņš 2015).

Already at the beginning of the Soviet occupation, tourism was used as a means of Soviet propaganda in serving both local and foreign tourists in Soviet Latvia. Tourism and excursion management and supervision institutions were established in the LSSR (Latvian Soviet Socialist Republic), the names of which were changed frequently, but the official tasks were quite unchanged: to organize tourism and excursions within the republic, including propagating Soviet ideological guidelines, acquainting workers with Soviet history, culture, economy, geography, and natural resources; to train tourism workers to manage tourist clubs; to build, maintain, manage, and finance tourism infrastructure; to organize trips and excursions, as well as to prepare tourists for receiving "USSR tourist" badges; to organize and implement consultations on tourism and excursion issues (Vaivode 2020).

An Attestation and Qualification Commission was set up to determine the training of tour guides and award the relevant qualifications. They decided that all tour guides hired for the first time must pass an attestation, where they were awarded a qualification with a certificate valid for three years. At the end of 1961, the Soviet authorities in charge of tourism decided to strictly limit the self-expression of tour guides during tours—the administration had to create control texts and a list of recommended literature for each route (Vaivode 2020).

In 1961, the Riga Excursion Base was established under the auspices of the Latvian Tourism and Excursions Board, which had to organize excursions around the republic. The Riga Excursion Base broadly expanded the development of tourism in Riga by developing 12 new route projects around Riga, paying attention to the promotion of the achievements of the republic's workers, the development of Riga in the fields of industry, housing, science, culture, and everyday life. In May 1962, new tourist routes had already been established in Riga, including revolutionary monuments, revolutionary events, the struggle of the Communist youth for Soviet power in Latvia, as well as literature, theater, and architecture. Quality control was to be performed every month, the narration of guides had to be listened to, inappropriate information could not be disseminated, and tourist routes and the guide profession had to be ideologized (Vaivode 2020).

The situation regarding the organization of foreign tourism was even more complicated. Restrictions on the movement of foreign tourists in the USSR were fixed in a 64-page set of regulations issued in 1936. The main point was that any movement beyond the "Intourist"-routes were prohibited. At the end of the Second World War, when the Soviet army occupied Latvia, similar processes were replicated in Soviet Latvia. After Stalin's death in 1953, the political regime was gradually liberalized, and the issue of admitting foreign tourists became important. In the mid-1960s, when the rapidly growing dynamics of foreign tourists became visible, the authorities established the operational management system of the Riga branch of SJSC "Intourist," which organized foreign tourism and operated until the end of the 1980s. This system guaranteed the conditions for the KGB employees in the institutions related to the reception and service of foreign tourists to freely recruit and cooperate with the existing agents and confidants, to receive their reports, and to track foreigners. One of the preconditions for such activities was to ensure that people who supported the interests of the KGB and met the requirements were appointed to management positions (Lipša 2017).

After each tour, the guides had to make notes in so-called workbooks, starting with the recording of their observations and ending with recounting the questions asked by the tourists and their conversations. The records were supposed to be a secret; they had to be kept in a safe, and the notebooks were destroyed over time by the KGB as secret records. Managers of these guides prepared reports on the work of the department based on the notes. Most likely, the records of the guides were also read by the officer who curated the foreign tourism department in the republic from the KGB (Lipša 2017, p. 87).

Needless to say, the routes developed and controlled by Soviet and foreign tourists department did not include anything about Holocaust memorials in Riga. In addition, foreign tourists were very strictly controlled, and it was almost impossible to get to the sites of their choice. The tourist guides published in all languages also did not offer information on such topics, nor was it available on public maps. The first unofficial guide, "Ebreju Rīga" [Jewish Riga] (1990) by Holocaust survivor and historian Margers Vestermanis, became a guidebook for the first local and foreign travelers interested in the difficult historical heritage, and especially Jewish history. And for many years, this guidebook was the main and only textbook for guides, who began to learn to lead tours of Riga's Holocaust memorial sites through self-education.

3   The Darkest Sites of Dark Tourism in Riga—Sites Related to Jewish Heritage and the Holocaust

After the fall of the Iron Curtain and the restoration of the country's independence in 1991, Latvia, Riga, and the places associated with the tragic events of WWII and the Holocaust became accessible to foreigners as well. Today, museums, like "Jews in Latvia," Museum of the Riga Ghetto and Holocaust in Latvia,

and memorials, like Riga Choral Synagogue, Jewish Memorial at Rumbula, Jewish Memorial Bikernieki, and Žanis Lipke memorial have become an integral part
of Riga tourism. Information about these places is available on the Riga Tourism
website (www.liveriga.com). Museums and memorials linked to Jewish heritage
and the Holocaust are included in the category of dark tourism sites related to
death, disaster, and suffering (Lenon and Foley 1999). Stone defines dark tourism
as "travel to sites of or sites associated with death and 'difficult heritage' within
global visitor economies" (Hartmann et al. 2018, p. 279). According to Tarlow
(2005, p. 48), these "tragedies or historically noteworthy death continue to impact
our lives." According to Stone (2006), Holocaust sites are among the "darkest"
spectrum of dark tourism, and their characteristic product features are education, conservation and commemoration, and authenticity of product interpretation.
There is a difference between sites that are of death and suffering, where disasters happened (primary sites), and sites created in other locations associated with
death and suffering (secondary sites) (Lennon and Foley 1999; Miles 2002; Wight
2006), or *in populo sites* "which embody and emphasise the story of the people
to whom the tragedy befell" (Cohen 2011, p. 194).

In the case of Riga, official tourist information is available for the primary
sites Riga Choral Synagogue, Jewish Memorial at Rumbula, Jewish Memorial
at Biķernieki, the secondary sites Riga Ghetto & Latvian Holocaust Museum,
Žanis Lipke memorial, and the *in populo* museum "Jews in Latvia." All of them
have become tourist attractions. It should be noted that Holocaust-related sites
are in no way directly comparable to other tourism "products" (Beech 2000).
According to Stone's seven dark suppliers framework (Stone 2006, p. 152–157),
the category of "darkest sites" includes "Dark Exhibitions, Dark Resting Places
and Dark Camps of Genocide."

The category of Dark Exhibitions comprises dark tourism "products," such
as museums, whose primary function is commemoration and education. These
attractions are often not located at the place of death (or a tragic event) and have
a slightly more developed tourism infrastructure and commercialized approach
(Stone 2006). For Riga, this group includes museums that directly reflect the
tragic events and the Holocaust in Latvia, and the Žanis Lipke Memorial, dedicated to the memory of the Jewish savior Žanis Lipke. In the last decade,
museums commemorating Jewish saviors have opened in many European countries, such as France, Germany, Poland, Lithuania, Bulgaria, and other countries
(Wóycicka 2019).

The category Dark Resting Places includes cemeteries or grave markers whose
"key product features revolve around a history-centric, conservational and commemorative ethic" (Stone 2006, p. 155). The Old Jewish Cemetery, where a park

was established in a closed cemetery during the USSR, is an example for this category.

Following Stone, the darkest of the darkest sites are the Dark Camps of Genocide, located in places where macabre events, atrocity, and genocide took place. Their primary focus is memory and education; these places often offer limited historical knowledge, and they are "produced to provide the ultimate emotional experience" (Stone 2006, p. 157). These sites are not designed to attract tourists, so tourism infrastructure is often limited.

The Jewish heritage and the Holocaust sites included in Riga's official tourism information are only some of the sites related to Jewish heritage and the Holocaust. Sites such as the Riga Ghetto, the concentration camps Kaiserwald, Dunamunde, and others, are left out of official information.

While tourism information refers only to a part of the sites related to the history of the Jewish community and the Holocaust, there has been a growing demand for these places over the last decade, until the Covid-19 pandemic in 2020. According to the statistics compiled by museums, the total number of visitors to Riga Dark Exhibition sites has increased over the last ten years (Table 1). The Museum of the Occupation of Latvia (Latvian Cultural Data Portal 2021) is an exception, where a small part of the exposition is dedicated to the history of the local Jewish community. The number of visitors to the Riga Ghetto and Latvian Holocaust Museum (2021) has tripled since 2013, reaching 25,644 visitors in 2019. A similarly rapid increase in the number of visitors is also observed in the Žanis Lipke memorial (2021). The number of visitors to the museum "Jews in Latvia" (2021) increased by 1.6 times.

A significant number of Riga's museum visitors are foreign tourists. For example, data from the Riga Ghetto and Holocaust Museum reveal that in the period from 2013 to 2019, 26% of visitors were local and 74% were foreign. Of the total number of visitors, 18% came from Germany, 16% from Great Britain, 11% from the US, 9% from Israel, 7% from Russia, 4% from France, and 3% from Finland, the Netherlands, and other countries. According to the information for 2019 of the museum "Jews in Latvia" (2021b), one third of the visitors visited the museum in organized groups, and 57% of the visitors in these groups had come from abroad. The trend of Žanis Lipke Memorial (data provided by the museum) in recent years has been that the share of foreign guests in the tourist season is about 50%, and in 2019, 34% of visitors who arrived in organized groups were foreign tourists.

The motivation of visitors for visiting dark sites are different. Significant motives for visitors are "desire for education, learning, understanding about what happened at the dark site, cultural and social reasons, etc." (Iliev 2020, p. 17).

Table 1 Visitor numbers in Riga's sites related to Jewish heritage and Holocaust (2010–2020): data provided by the museum (*) and by the Latvian Cultural Data Portal (**)

	2010	2011	2012	2013	2014	2015	2016	2017	2018	2019	2020
Museum "Jews in Latvia"*	4,335	5,348	5,013	5135	6,545	6,012	7,510	6,714	7,602	8,152	1,723
Occupation Museum**	24,2954	10,3249	86,392	72,252	136,045	122,366	102,368	11,6631	103,559	85,775	27,677
Riga Ghetto and Holocaust Museum*	–	–	–	8,372	17,317	21,304	27,518	20,009	20,121	25,644	5,928
Žanis Lipke Memorial*	–	–	–	3,415	7,026	8,164	7,651	8,104	10,344	10,947	3,645

Biran et al. (2011, p. 836) divide tourists into three groups according to the motivations for visiting the darkest sites, which "include a desire to learn and understand the history presented, a sense of 'see it to believe it' and interest in having an emotional heritage experience."

In Riga, groups with a personal connection to the site who perceive the site as a personal heritage and want to gain an emotional experience are mainly travelers of Jewish descent from Israel and the United States. Their reasons are related to the search for family roots, the commemoration of deceased relatives, or the common Jewish tragedy. These groups or independent travelers often use the services of travel companies, and the route can include only Riga or the territory of Latvia and neighboring Lithuania. Jewish Riga guided tours are integrated in Baltic Sea cruises. Travel companies cooperate in organizing these thematic tours with guides who specialize in this field and with whom successful cooperation has been established, but at the same time admit that there are a small number of Hebrew-speaking guides in Riga, so there are cases when Baltic tours have to use guides from Lithuania.

A significant number of tourists express their desire to acquire knowledge and understand historical events (Biran et al. 2011) as part of a leisure trip. This motive applies to the darkest sites and other heritage sites (Timothy and Boyd 2003). The increase in the number of tourists in Riga goes hand in hand with rising numbers of visitors to places related to Jewish heritage and the Holocaust. Both independent tourists and organized groups visited these sites, especially from Germany, the United Kingdom, Russia, Sweden, and Finland. The guide plays an important role in the co-creation of the experience of these groups. Specialized tours are provided in Latvian, English, German, and Russian, but there is limited availability of Jewish Riga tours in other languages.

Data on the number of visitors to the Dark Camps of Genocide are not available, given the memorial function of these places, but these places have also become important tourist attractions. They are included in guided tours for both local and foreign groups of travelers, as well as in the routes of independent travelers. For example, the Biķernieki Memorials and the Rumbula Forest Memorial are described in foreign traveler reviews on the international travel portal TripAdvisor (TripAdvisor 2021) as "an important visit," "humanity's darkest site," "stark and sombre," "the most moving Holocaust memorial I've seen," "powerful memorial," and "a memorial that stuns you into silence and sorrow." Interestingly, visitor reviews also point to the underdeveloped tourism infrastructure in these places.

4   Experiences in Guiding Jewish Riga

In April and May 2021, the experience of Riga tourist guides who directly or indirectly worked on the Holocaust memorial site route or its separate objects in Riga was clarified through qualitative research. In this study, purposive sampling was used. Due to Covid-19 restrictions on meeting people face to face, structured interviews in written form for primary data collection were distributed to the guides through the Jewish memorial sites (museums and memorials) in Riga. These institutions spread via email the interview questions to the guides with whom they work most frequently. Fourteen tourist guides responded to the interview. Anonymously or publicly, according to their free choice, participants answered ten questions related to the experiences and challenges of their work (professional and human) and educational opportunities. An inductive approach was used for qualitative data analyses, and data was analyzed manually. All quoted comments are translated from Latvian to English. All participants are identified by code name to protect their privacy. The group of participants includes tourist guides who started their careers shortly before the collapse of the Soviet system or who just got involved in new professional projects. There were six men and eight women of various ages and at different stages in their professional lives, with higher education diplomas in the humanities and even some doctoral degrees. Either an undergraduate or graduate education in history, theology, pedagogy, or the acquired professional qualification of a guide in various professional courses, enabled them to work as professional guides. One guide has a higher education degree in theology with a specialization in Holocaust history.

All guides indicated that at the same time, they supplemented their knowledge in various Latvian and German museums and courses organized by TIC or tourism companies. Only one guide referred to "Intourist" courses attended during the USSR. History as a hobby, self-education and information obtained from various sources, as well as literature and periodicals were indicated as important sources for acquiring the essential knowledge necessary for a guide in modern day Latvia.

Becoming a guide in post-Soviet and communist countries since the 80s of the twentieth century was not an easy task. The previous ideological training became redundant, and the knowledge was insufficient or invalid. With the collapse of the old Soviet system of organizing and monitoring tourism, many guides lost their jobs, were forced to retrain or, at best, learn new foreign languages that had not been needed in the Soviet tourism system, and had to join the field of outbound tourism. In addition, the inhabitants of Latvia had unprecedented opportunities to travel. Communication and sociology expert Bērziņš believes that in 2000 a significant part of the Latvian population knew something more about the Holocaust in the world than previously in Latvia (LSM 2021). A large part of family and individual savings was set aside for foreign travel, domestic tourism figures in Latvia declined sharply, but the offers of tourist routes increased rapidly. Free travel attracted many tourists of Latvian Jewish origin from all over the world. Since the relatives and friends of most Jews of Latvian origin were killed in the Rumbula woods in Riga, it became one of the main stopping points. The Biķernieki Woods, on the other hand, was the most important stopover for visiting kin of deported Jewish relatives from Germany, Austria, and Czechoslovakia. Holocaust memorial site routes in Riga and Latvia were a new and completely unknown world for many to discover and explore. In parallel with opening up of discussions on previously forbidden topics such as the Second World War and the consequences of the Nazi and Soviet occupation of Latvia, issues regarding collaboration of the inhabitants and the topic of the Holocaust became some of the most important objects of historical discussion and research. A long time has passed since the end of the Second World War and the tragedies of the Holocaust that destroyed most of Latvia's Jewish population. Much had been forgotten, hidden, silenced, and lost. As Melers, a researcher and explorer of Jewish cemeteries and Holocaust sites, wrote in his book *History of the Latvian Jewish Community and Holocaust Memorials*: "Unfortunately, after more than six decades, when two or three generations have changed, in some areas of Latvia where 20–30 to 50% of the population were Jews, the current population has often never seen any Jews in their lifetime and knows nothing about them" (Melers 2013).

Riga was in a rather privileged situation, because there was a large and active Jewish community with synagogues and schools. Also, the historical memory of the Jews before the war had not completely disappeared, and came to life over time. As one of the guides has pointed out in his experience story, local tour participants of the older generation (not Jews) have become more open and willing to share their experiences, impressions, and emotions. These changes in the "openness" of the local community are certainly related to the influence of savior themes (narratives): "people themselves begin to share their stories about their Jewish neighbors [...] I often encounter moments of silence—people are overwhelmed by emotions. [...] the older generation tends to admit that they were happy to hear stories of Latvians saving Jews rather than those of killing them" (G9).

Part of the older pre-war generation of Latvian Jews became tourist guides, showing visitors the Jewish heritage of Riga. A guide with extensive experience working on this route for several decades pointed out an important fact in his questionnaire: "When I started my guiding career (30 years ago) and started offering tours of Jewish places in Riga, no one knew about them; now they know" (G4).

Later, in the early 2000s, they were replaced by the younger generation of a very few Jewish guides in Riga, who took over knowledge and experience from the elders and developed their professional activities. Some of them emigrated to Israel and developed their tourism business between Latvia and Israel. Quite often, Jewish emigrants from the former USSR were the first foreign guides to lead excursions in Latvia. In addition to the activities of Jewish tourist guides and the specialized Jewish routes they created in Riga, a new generation of non-Jewish professionals emerged, who, alongside general Riga tourist routes, began to include Holocaust memorials. Here, one of the greatest challenges of the 1990s and early 2000s was the lack of knowledge about the history of Latvia, the Jewish community, and the Holocaust; even the academy was not able to provide information.

A new generation of tourist guides entered the new free tourism market and the capitalistic system—young people who had language skills but no relevant education. During the Soviet period, tour guides were mostly people with education in humanities, whereas after 1991, most of the guides' education and professional development was left to their own choice. The knowledge of the newly educated guides is much broader and the level of discussion much deeper than that of the Soviet-educated ones. They do acknowledge this broader perspective: "The new generation understands events better than those who studied during the Soviet era, [which was] strongly influenced by propaganda" (G2).

Another guide stated that professional tour guide organizations were quite unpopular or politically uninfluential. At the institutional level the self-organization or solidarity of tour guides was so low to change anything in the professional field of tour guides. However, individual exchanges of experience, information, and knowledge and best practices among tourism guides were much more effective than institutional cooperation (G13).

In the 1990s, the Latvian government gradually improved its support for evaluating history. The state promoted historical research and educational programs, as well as the preservation of memory. This work became more active in the second half of the 1990s, and interest in cooperating with the International Working Group on Holocaust Education, Research and Acceptance emerged. On February 7, 2001, the museum and documentation center "Jews in Latvia" was granted the status of a state-accredited museum. Its state accreditation demonstrated the political stance of the Latvian state regarding research on the past, including research on the history of the Holocaust and minorities. Following its accreditation, the

Latvian government provided funding to the center "Jews in Latvia" to prepare an exhibition on the history of the Jewish community and the Holocaust in Latvia. The subject of the Holocaust was included in the compulsory history curriculum as a part of general education. In the autumn of 1998, the President of Latvia, Guntis Ulmanis, established the International Commission of Latvian Historians. Its aim was to promote research issues concerning Latvia in the twentieth century and to explain it to the society in Latvia and other countries. In its work, the Commission paid special attention to two totalitarian regimes that followed one another and committed crimes against humanity in the territory in Latvia. In October 2000, the Commission of Latvian Historians organized the first international conference on the problems of Holocaust research in Latvia. The research was published in a separate volume, thus providing the general public with an insight into the latest scientific research (Dribins et al. 2001). These academic and educational projects developed in cooperation between public and private institutions have contributed to the discourse on Jewish history, both in Latvia and internationally: "The biggest change has been in public consciousness—the information vacuum about the history of the Jewish community, outstanding personalities, and tragic death has been filled, and a discussion is taking place on controversial issues" (G10).

Conferences and collections of articles both became periodic events, and their main audience was educators and museum professionals. However, none of the institutions provides supportive measures to the professional activities of a tourist guide, for example, professional ethics, intercultural communication, or methodology. Further, they did not reflect on guiding difficult historical heritage from a tourist and economic perspective. Knowledge related to psychology and communication is also often underestimated, and no institution in the network helps tour guides in the process of their professional development, although the guides themselves acknowledge that this is one of the biggest challenges in their work: "controlling their own emotions" (G9) during the tour, when narrating about tragic events; their narrating to "rescued Jews about their deceased relatives" (G9) creates emotionally difficult moments for both Riga guides and tourists. Challenges related to tourist behavior are, on the one hand, how to "gain empathy" (G5) in the participants about the tragedy of the local Jewish community or how to deal with the "sensitivity of relatives of Holocaust victims" (G8). One of the most important professional tools for guiding on Holocaust and Jewish history routes is knowledge of both intercultural communication and Jews in general, and the Jewish history. One of the Riga guides stressed this knowledge as one of the most important tasks. He believed it was important to know the history of the Latvian Jewish community and the tragedy of the Holocaust, and also to know the global history of Judaism, Jews, and the country from which these guests come from:

> "People are so different politically, religiously, culturally, emotionally [...] The ability to listen and hear tourists is even more important than the story itself. Quite often, when I started my job, I was confronted with the fear of non-Jewish guides working with Jewish tourists or on routes, especially forged by guides from the Soviet system, related to the history of the Holocaust. And not because they had prejudices against Jews or Judaism, but because they were afraid of questions, they were afraid of making mistakes. The absurdity of the Soviet system was that guides had rehearsed questions and answers and did not think outside the box. However, in post-Soviet life there was no knowledge of different cultures, religions and their differences. When I told them not to be afraid of questions posed by Jewish tourists, because they are not affronts, but a living culture of discussion keeping in with the thousands of years of Jewish traditions, they were both confused and thoughtful" (G11).

Mcdonell stresses that guides are a cultural bridge, and they must be experts in the culture of destination and have a good understanding of the culture of tourists (Mcdonell 2001).

Guides had acquired and supplemented their knowledge of Jewish heritage in Riga and the history of the Holocaust through a wide variety of sources, in particular, courses offered by the Museum "Jews in Latvia" and exhibits at other museums (including the Žanis Lipke Memorial and the Riga Ghetto and Latvian Holocaust Museum); books and articles on the history of Jews and the Holocaust (Bērziņš 2017); stories of memories read or listened to; experience in museum work; participation in lectures, conferences, and seminars (including events organized by the University of Latvia Judaica Centre, the Museum "Jews in Latvia," the Riga Ghetto and Latvian Holocaust Museum, the Žanis Lipke Memorial); films, and other sources. Guides had also supplemented their knowledge outside Latvia, for example, attending lectures and seminars organized by the Yad Vashem International School of Holocaust Studies or by visiting Jewish Holocaust sites in Lithuania, Poland, and Germany.

Latvian guides in their professional field are very vulnerable both from the influence of the tourism business and money, as well as from the current international political situation. Local Riga guides have experienced situations when tourists from Russia "do not want to believe the story of historical events in Latvia" (G2), when there is "radicalism, ignorance, or misinformation perceived by certain tourists as 100% truth" (G10).

Speaking of mutual cooperation between tour guides and tour operators, over the last 30 years in Latvia there has been a rather elementary and unsustainable cooperation, where buyers of services have had no interest in developing cooperation or investing in professional development of guides or wider public projects, which would increase the visibility and popularity of Holocaust memorial sites. As highlighted by one of the respondents:

> "Working in this field for 15 years, I am convinced of one distinct feature of the tourism business environment in today's world. Tour operators are not so much interested in quality as in the possibility of getting cheaper guide services and earning more. Rarely, when talking about the legacy of difficult history, was there an insistent request to emphasize the scale of the tragedy, to tell the darkest stories" (G11).

According to the survey, guides gained knowledge from conferences or organized seminars:

> "[B]ack in the early 2000s, despite my academic education as a historian, my knowledge of Jewish history and the Holocaust was acquired from conversations with Marģers Vestermanis, Meijers Melers, Zāmuels Etelsons, and other members of the Riga Jewish community, especially Holocaust survivors, and conferences organized by the Presidential Commission in cooperation with the Latvian Jewish community and expositions offered by the museum 'Jews in Latvia' as well" (G11).

Several guides emphasize that a large number of Riga's tourist guides and Holocaust memorials acquired their knowledge directly from the Jews who survived the Holocaust, who selflessly shared their knowledge and told their personal stories, all of which helped to fill the gap in knowledge and strongly influenced professional development of guides more than academic books.

Guides have experience working with a very wide and very diversely motivated audience. The interest in a tour and excursion has been very different, ranging from "formal (interests) or common story" (G10), to the "traditional" tourists of Riga, whose excursion choice had been to get to know the culture and history of Riga (heritage tourism). These results match those observed in earlier studies by Timothy and Boyd (2003) and Biran et al. (2011). As one of the guides pointed out: "The Jewish theme is inseparable from the Riga tour" (G1). This group of travelers includes both independent as well as group travelers from Latvia and abroad, like Germany, Russia, Sweden, Great Britain, the Netherlands, Denmark, Poland, France, Australia, South Africa, and many other countries.

A significant role in tourism is played by groups of local high school students, students and seniors, as well as groups of students and young people from abroad, mainly from Germany. Tour guides who collaborate with German tour operators, German research institutes, and German twin cities acknowledge that "this is the largest group of travelers looking for the footprints of Jewish citizens of Riga and showing a deep interest in Jewish history" (G14). In addition, German-speaking guides quite often take part in educational seminars on the history of the Holocaust in Germany, and they visit German Holocaust monuments and memorial sites. Contacts and cooperation established with German guides are a professionally important tool for development and a way to enrich one's knowledge (G14).

A large group of travelers is interested in "personal family history" (G10), often represented by Jewish community travelers from Israel, the United States, and other countries. Their motive for travel to Latvia is to find one's family roots (for example, finding a birthplace) or visiting the memorial sites of relatives who died in the Holocaust. As one of the guides notes (15 years of experience with this group): "I have helped to obtain archive materials, compile family trees, and find ancestral properties" (G10). These results reflect those of Biran et al. (2011), who found that a significant group to the darkest sites are people with a personal connection to the site.

All the Riga guides surveyed indicated that, despite all their individual or professional challenges, they are satisfied with their profession (G4). Guides generally describe their experience as very positive, pointing out that the experience is "emotionally rich because it is associated with a sensitive history" (G5). Additionally, they stress the opportunity to satisfy the request of customers, supplement their personal knowledge, learn a lot of new things, and "learn about what we have lost with the disappearance of the Latvian Jewish community" (G10).

5   Conclusions

After the Second World War, the Soviet authorities in Latvia pursued a policy of concealing the Holocaust and its remembrance in the form of refusals and evasions to build monuments and erect memorial sites. The Jews of Riga who survived the Holocaust tried to identify the sites of the victims of the Holocaust and create a culture of remembrance for the dead. The first informal guides to commemorate events and visit Holocaust sites were mostly Holocaust survivors from Riga or their relatives, who in a way combined two responsibilities—amateur research and storytelling functions. Neither did the routes developed and controlled by the Soviets for foreign tourists include anything about Holocaust memorials in Riga, nor was it possible for independent tourists to get to the places of their choice, as the tourists were very strictly monitored and controlled.

With the collapse of the Soviet system of organizing and monitoring tourism, the previous ideological training became redundant, and the knowledge—insufficient or ineffective. In Riga, the culture of Holocaust memorials developed much more successfully and efficiently, thanks to the fact that the tradition was kept alive and active by the Jewish community, whose original backbone was formed by Jews who survived the Holocaust. Synagogues and Jewish schools functioned after 1991 as well. In the Soviet system, tour guides were mostly people with a humanitarian education and a strict ideological indoctrination. After 1991, most of the tour education and professional development was left to the people themselves, creating a certain gap between the "new and old" generation of tour guides.

The history of the Jewish community and the Holocaust has become an important part of Riga's heritage tourism. The demand for Jewish heritage and Holocaust sites is growing year by year, offering an experience to both travelers who have a personal connection to these sites and those for whom the Jewish experience of Riga is part of getting to know the common heritage. The declared aim is to gain knowledge and provide a better understanding of historical events.

Riga's tourism supply includes places where tragic events of the Holocaust took place and sites associated with tragedy. Riga's official tourism information is available on the "darkest sites" of dark tourism such as Dark Exhibitions, Dark Resting Places, Dark Camps of Genocide (Stone 2006), but this offer can be described as narrow and includes only the most popular or visible sites.

One of the most important tools for the professional work of a guide for the Holocaust and Jewish history is knowledge of both intercultural communication and general knowledge of Jews and Jewish history, which has not been fully covered in the process of guide education and professional development.

Guides are very vulnerable in their professional field to the impact of the tourism business and money, and the current international political situation. The current situation, where guides are mostly dependent on various professionals in the tourism business, is not sustainable and requires solutions, especially in areas of tourism such as difficult history and its heritage. It is possible that more active involvement of the community (in its broadest sense) could change this situation.

Riga's Jewish heritage and Holocaust itineraries offer tourists a broad experience and a wider range of sites. Without using the services of a travel company or tour guide, there is a high probability that these objects cannot be found, significantly affecting independent heritage tourist experience in Riga. Therefore, it would be important to highlight this section in the official tourist information on the historically important Jewish community in Riga by enriching information about sites included in guide routes related to traditional residences, Holocaust memorials, and notable personalities. It would also be important to identify the condition of the tourism infrastructure and information provided in these places, as well as any improvements that are necessary. Professional guides play an important role in educating tourists and building experience (Weiler and Black 2015), especially on difficult local heritage topics, so official tourism information should include information on guides' specialization topics.

Acknowledgements This research as a part of the project "Difficult Heritage: Between the Memorisation and Contemporary Tourism Production and Consumption. The Case of Holocaust Sites in Latvia" (MemoTours), project No. lzp-2019/1-0241 and is funded by the Latvian Council of Science.

References

Beech, J. (2000). The enigma of holocaust sites as tourist attractions-the case of Buchenwald. *Managing Leisure, 5*(1), 29–41.

Bērziņš, D. (2015). *Sociālās atmiņas ētika un komunikācija: holokausta diskursi Latvijā (1945–2014).* Rīga: LU Latvijas vēstures institūta apgāds [Ethics and Communication of Social Memory: Holocaust discourse in Latvia: 1945–2014].

Bērziņš, D. (2017). Holocaust Historiography in Latvia: The Road Toward Research Infrastructure. *Dapim. Studies on the Holocaust, 31*(3), 276–284, https://doi.org/10.1080/232 56249.2017.1395526.

Biran, A., Poria, Y., & Oren, G. (2011). Sought experiences at (dark) heritage sites. *Annals of tourism research, 38*(3), 820–841.

Cohen, E.H. (2011). Educational dark tourism at an in populo site: The Holocaust Museum in Jerusalem. *Annals of tourism research, 38*(1), 193–209.

Die Jüdische Gemeinde "Schamir" (2015). *Karte der Gedenkstätten der jüdischen Geschichte. Synagogen, Holocaust-Gedenkstätten, Jüdische Friedhöfe.* Riga: Latvijas Karte.

Dribins, L., Gūtmanis, A., & Vestermanis, M. (2001). *Latvijas ebreju kopiena: vēsture, traģēdija, atdzimšana.* Rīga: LU Latvijas vēstures institūta apgāds [Latvian Jewish community: history, tragedy, rebirth. Riga: LU Institute of Latvian History]. www.mfa. gov.lv/ministrija/publikacijas/latvijas-ebreju-kopiena-vesture-tragedija-atdzimsana#pil soni. Accessed 21 May 2021.

Druva-Druvaskalne, I., Abols, I., & Slara, A. (2006). Latvia Tourism: Decisive Factors and Tourism Development. In D. Hall, M. Smith, & B. Marciszewska (eds.), *Tourism in the New Europe: The Challenges and Opportunities of EU Enlargement* (170–182). Wallingford, Cambridge: CABI.

Džads, T. (2012). *Pārvērtēšana: Esejas par aizmirsto 20. gadsimtu.* Rīga: Dienas Grāmata, 20. [Revaluation: Essays on the Forgotten 20th Century. Riga: Dienas Gramata]

Hartmann, R., Lennon, J., Reynolds, D.P., Rice, A., Rosenbaum, A.T., & Stone, P.R. (2018). The history of dark tourism. *Journal of Tourism History, 10*(3), 269–295.

Iliev, D. (2020). Consumption, motivation and experience in dark tourism: a conceptual and critical analysis. *Tourism Geographies*, 1–22.

Lennon, J.J., & Foley, M. (1999). Interpretation of the unimaginable: The US Holocaust Memorial Museum, Washington, DC, and "dark tourism." *Journal of Travel Research*, 38(1), 46–50.

Light, D., Creţan, R., Voiculescu, S., & Jucu, I.S. (2020). Introduction: Changing Tourism in the Cities of Post-communist Central and Eastern Europe. *Journal of Balkan and Near Eastern Studies*, 22(4), 465–477.

Lipša I. (2017). VDK ietekme un padomju Latvijas ārzemju tūrisma iestāžu vadošais personāls: Vissavienības akciju sabiedrības "Intūrists" Rīgas nodaļa un tās operatīvā vadība (1957–1992). *Latvijas vēstures institūta Žurnāls*, 2(103), 80–128. [Influence of the KGB and the leading staff of Soviet foreign tourism institutions of Latvia: Riga branch of the SJSC "Intourist" and its operational management (1957–1992)].

LSM (2021). *Zinātnes vārdā. Holokausta piemiņas vietas PSRS radīja Rietumu spiediena dēļ. Saruna par ebreju memoriāliem.* www.lsm.lv/raksts/dzive--stils/vesture/holokausta-pieminas-vietas-psrs-radija-rietumu-spiediena-del-saruna-par-ebreju-memorialiem.a37 9965/. Accessed 21 May 2021.

Mcdonell, I. (2001). *The Role of the Tour Guide in Transferring the Cultural Understanding.* Sydney: School of Leisure, Sport and Tourism.

Melers, M. (2013). *Latvijas ebreju kopienas vēsture un holokausta piemiņas vietas.* Rīga: LU Filozofijas un socioloģijas institūts. [History of the Latvian Jewish Community and Holocaust Memorials].

Miles, W.F. (2002). Auschwitz: Museum interpretation and darker tourism. *Annals of tourism research*, 29(4), 1175–1178.

Museum "Jews in Latvia" (2021a). *Tracing Roots in Latvia.* http://jewishmuseum.lv/lv/item/136-riga.html. Accessed 13 June 2021.

Museum "Jews in Latvia" (2021b). *Visitor statistics 2010–2020* (unpublished data).

Riga Ghetto and Holocaust Museum in Latvia (2021). *Visitor statistics 2010–2020* (unpublished data).

Rozite, M., & Klepers, A. (2012). Out of the Soviet Union: The re-emergence of Rīga as a tourist capital. *Current Issues in Tourism*, 15(1-2), 61–73.

Stone, P.R. (2006). A dark tourism spectrum: Towards a typology of death and macabre related tourist sites, attractions and exhibitions. *Tourism: An International Interdisciplinary Journal*, 54(2), 145–160.

Tarlow, P. (2005). Dark tourism—the appealing "dark" side of tourism and more. In M. Novelli (ed.), *Niche Tourism: Contemporary issues, trends and cases* (47-58). Oxford: Elsevier Butterworth-Heinemann.

The Investment and Tourism Development Agency of Riga (2021a). *Rīgas tūristu gidu sertifikācija (in Latvian).* www.liveriga.com/lv/7488-rigas-turistu-gidu-sertifikacija. Accessed 5 June 2021.

The Investment and Tourism Development Agency of Riga (2021b). *List of certified Guides in Riga.* www.liveriga.com/en/7489-list-of-certified-guides-in-riga. Accessed 5 June 2021.

Timothy, D. J., & Boyd, S.W. (2003). *Heritage tourism.* Edinburgh: Pearson Education.

Vaivode, I. (2020). Tūrisma organizācija Latvijas PSR: Latvijas republikāniskās tūrisma padomes darbība 20. gadsimta 60. Gados. *Latvijas Universitātes žurnāls. Vēsture*, 9–10. [Tourism organization Latvian SSR: Activities of the Latvian Republican Tourism Council in the 1960s] https://doi.org/10.22364/luzv.9.10.05. Accessed 13 June 2021.

Weiler, B., & Black, R. (2015). *Tour Guiding Research: Insights, Issues and Implications*. Bristol: Channel View.

Wight, A. C. (2006). Philosophical and methodological praxes in dark tourism: Controversy, contention and the evolving paradigm. *Journal of Vacation Marketing*, 12(2), 119–129.

Wóycicka, Z. (2019). Global patterns, local interpretations: new Polish museums dedicated to the rescue of Jews during the Holocaust. *Holocaust Studies*, 25(3), 248–272.

Žanis Lipke Memorial (2021). *Visitor statistics 2010–2020* (unpublished data).

Links

Lonely Planet (2021). *Latvia. Riga. Top things to do*. www.lonelyplanet.com/latvia/riga/top-things-to-do. Accessed 21 May 2021.

Latvian Cultural Data Portal (2021). *Statistical data of Museum of the Occupation of Latvia*. https://kulturasdati.lv/lv/muzeji/latvijas-okupacijas-muzejs. Accessed 21 May 2021.

Tripadvisor (2021). *Bikernieki Memorial*. www.tripadvisor.com/Attraction_Review-g27 4967-d4813337-Reviews-Bikernieki_Memorial-Riga_Riga_Region.html#REVIEWS. Accessed 13 June 2021.

Israel

Identificatory Trajectories of Holocaust Memorial Site Guides

A Theory Sketch Grounded in Yishai Sarid's Novella *The Memory Monster*

Franz Breuer

"I think that in order to survive we need to be a little bit Nazi, too, he said" (Sarid 2020, p. 125).

"We have to be a little bit Nazi. You've finally said it. You got the point, kids, well done" (Sarid 2020, p. 127).

Abstract

Yishai Sarid's novella *Mifletzet HaZikaron* (*The Memory Monster* 2020) depicts the experiences of a young Israeli historian who guides tour groups—mainly Israeli high school students—through the former Nazi German concentration and extermination camps in Poland. Adopting a grounded theory research approach, the present article provides a theoretical sketch of the identificatory stances of such memorial site guides and how these change over time. The report of the protagonist in Sarid's text (in the English translation) is used as data source for developing this draft. First, the concept of personal (un)connectedness with (Holocaust) memorial work is worked out. Second, a trajectory model is proposed in which personal control over the world of action and control of events by overwhelming circumstances are placed in a

I thank Miriam Geoghegan for translating this chapter into English.

F. Breuer (✉)
Westfälische Wilhelms-Universität Münster, Münster, Germany
e-mail: breuerf@uni-muenster.de

dynamic relationship to each other. Because the theoretical sketch is undertaken as an individual case analysis, it requires differentiation and saturation
through further cases and data sources.

1   The Plot and Appeal of the Novella

The protagonist and first-person narrator in Yishai Sarid's novella *Mifletzet
HaZikaron* (English: *The Memory Monster* 2020; German: *Monster* 2019), is a
young Israeli historian with a wife and a preschool child. The story recounted in
the novella takes place in the years before and after 2010. The narrator initially
studies history and international relations, but after failing to gain admission to
the Israeli Foreign Service in his second year at university, he sets his sights on
working as a historian—preferably in the field of early history, in order to avoid
the turbulences and catastrophes of the Jewish people (Sarid 2020, p. 6). However, one of the few promising ways of making a living as a historian in Israel
is to specialize in Holocaust studies. Thus, after obtaining a small scholarship,
he does a doctorate on the logistics of the extermination machinery in the Nazi
German death camps in Poland. His doctoral thesis is meticulously researched,
and replete with horrific and macabre details. While writing the thesis, he works
on the site as a guide at Yad Vashem, the World Holocaust Remembrance Center,
in Jerusalem. During this time, he takes part in a Yad Vashem course in Poland
for group guides, and after receiving his license he begins to work as a guide
for groups of Israeli high school students on their educational trips to Holocaust
memorial sites in Poland.

Initially, Sarid's protagonist adopts the stance of an academically distanced
historian and lecturer, displaying all the competence and professionalism prescribed by the official canon of Israeli memorial site pedagogy. As he is able
to offer extremely detailed specialist knowledge, his guided tours are very well
received, and demand for his services grows. On one occasion, the management
of Yad Vashem hires him as an expert for extermination camps in Poland to
act as a consultant for the selection of the location for, and the planning of, an
act of commemoration of the 75th anniversary of the Wannsee Conference at an
extermination camp memorial site in Poland.

The narrator's work as a tour guide and consultant calls for long stays in
Poland—far away from his wife and young son, whom he gets to see only on
short visits to Israel. The family suffers under this balancing act. A problem arises
with the "fatherless" son; the child is being systematically bullied at kindergarten.
On a home visit, the narrator attempts to solve the problem in a somewhat brutal

way: he accompanies the child to kindergarten and menacingly threatens one of the bullies. The narrator and his wife unsuccessfully explore ways of remedying the geographical separation. Initially, the necessity of securing the family income, and later the lure of attractive career prospects in the Holocaust sector in Israel, militate against this.

With time, the academic distance and professionalism that the narrator initially displays in his work as a guide becomes fragile and increasingly disintegrates. The diversity of his social worlds and levels of identity—academic observer and professional mediator of historical facts and relationships; involved Jew (whose family history is not, however, revealed in the book); husband and father; empathetic person gifted with reflexivity—loses its equilibrium. His pedagogical memorial site praxis is thrown into identificatory and performative turmoil. Numerous unsettling moments that arise in the context of his work—in experiences with himself and others—render more and more difficult the canonically framed and interactionally cultivated exchanges with high school students and other visitors or clients (e.g., Israeli politicians and civil servants, members of the Israeli armed forces, a German film director) and bring the narrator to the limits of his mental and physical capacity.

In this state of emotional chargedness, things finally come to a head. On the recommendation of the chairman of Yad Vashem, he is hired by a German film director, who deceives him into thinking that he is merely acting as a guide through the Nazi extermination camps in Poland as part of his research for a film about the camps. In fact, the director films him with the intention of using the footage in the documentary. When the narrator realizes the full extent of the deception and the nature of the director's hidden agenda he explodes with anger and forcefully punches the man several times in the face. Although it is not revealed whether the assault has disciplinary consequences, we learn at the beginning of the book that the narrator has been requested to submit a report to the chairman of Yad Vashem accounting for his behavior. Yishai Sarid's novella thus takes the form of a letter to the chairman, in which the narrator reports how and why the outburst of affect came about.

Sarid's novella is a revelation in terms of the range of experiences, of the internal worlds and their dynamics in the pedagogical work of Holocaust memorial site guides. Their task is to accompany—to guide—today's high school students (the generation of smartphone owners and computer gamers) and their teachers, as well as other visitors (tourists, politicians, etc.) with the most diverse interests and motivations through the sites of the Nazi German persecution and extermination machinery in Poland and elsewhere. When doing so, they are expected to fulfil a number of specified pedagogical-didactic objectives.

Viewed through my German eyes, Yishai Sarid, as a Jew and an Israeli, can afford to ignore the reservations and (self)censorship mechanisms of the canon of political correctness in this domain. He accesses even the darkest psychological depths of observation, self-resonances, and associative reminiscences of his protagonist—an access that I can hardly imagine a German author achieving in an analogous text, be it of literary, cultural science, or social science provenance. Sarid broaches a wide range of subtexts of the experience of memorial sites and remembrance. He makes explicit what is frequently only whispered or said off the record. Moreover, he articulates things whose articulation violates taboos. Many Israeli readers will likely also feel this way. Here—in literary form—experiences of a memorial site guide are drawn upon that, in my view, bear extraordinary witness to the willingness of the author—or the first-person narrator—to reveal and reflect. A chapter in the novel *Quasikristalle* by the Austrian author Eva Menasse (2014) comes close to Sarid's text. In that chapter, the author describes the perspective of a memorial site guide at Auschwitz that is akin to that of Sarid's protagonist. At several points in what follows, I shall therefore draw also on Eva Menasse's text.

2   My Methodological Approach—Reflexive Grounded Theory

I approach Yishai Sarid's novella from a social scientific hermeneutic perspective that is informed by the methodology and methods of grounded theory (Glaser and Strauss 1967; Strauss 1987; Charmaz 2014) and reflexive grounded theory (Breuer et al. 2019). This research approach guides one to generate in a methodologically bottom-up way a new, innovative theory sketch for a delimited subject area. There is no question that this methodological discovery approach is indicated in the case of social science exploration and theory formation on the work of guides at Holocaust memorial sites. There is a dearth of relevant research on the (occupational) biographical trajectories of these guides, on their motivational engagement, their manners of personal self-involvement, their ways of dealing with stress, etc. This lack of research is also an indication of the underestimation of their importance for the mediation process. As Ballis (2018, p. 23) noted, "guides are rarely heard or interviewed," yet they "represent a crucial point of contact between the institutions and the visitors, between the past and the present, and between the space and the subject." Ballis and her research group, in particular, are endeavoring to redress this lack of research interest in the German-speaking area (Ballis 2018, 2019; Ballis and Gloe 2019).

In terms of methodology and methods, the empirical data of the research field take center stage in the grounded theory research approach. Through intensive hermeneutic engagement with these data (e.g., using specific types of coding), generalizing concepts (categories) are developed, which are then systematically related to each other (modeling). Early books on the grounded theory approach (Glaser and Strauss 1967, p. 21 f.) stated—in a metaphorical (and epistemologically problematic) manner of speaking—that a theory emerges from the data. However, this metaphor does not adequately take into account the active role of the human epistemological subject. Naturally, the data do not move on their own accord. Rather, intellectual midwifery on the part of the researcher is required to make them talk theoretically, as it were.

In this methodological context, data can be of all possible types and provenances. Ethnographic surveys, participant observation, and interviews or conversations with affected persons, participants, or field members define the style of the grounded theory approach. There are, however, no fundamental limitations in this regard. The principle that "all is data" (Glaser 2005) applies.

Hence, I use Sarid's novella as a data source and an information basis for a theory sketch in the same way as I would use information of this kind about a problem field or an account of biographical development from an interview with a field member (here: a memorial site guide). This type of heuristic use of literary texts has a certain tradition in grounded theory (Strauss 1987, p. 241 f.; Breuer 2009; Dieris 2009). It can, on the one hand, be justified with reference to the notion of the special psychological profundity, sharpness, and linguistic elaborateness of literary texts. On the other hand, it must be borne in mind that a literarization is not an account of experience in the strict sense, but rather that fictionality, accentuations, condensation, etc. play a role. However, there is never a one-to-one relationship between experience and report in the usual social science data (from interviews, participant observations, etc.) either. Rather, we are always dealing with communication addressed to a recipient that is shaped by the subjective filters and perspectives of the observers and intentional informants.

One constitutive epistemological tool of the grounded theory methodology is the comparison of different phenomena or variants (text passages, participant perspectives, data types, etc.). Deliberations on the selection of suitable cases for the development of a theory are of great importance. The concept of theoretical sampling—that is, "sampling aimed toward theory construction [...] not for population representativeness" (Charmaz 2014, p. 8)—plays a central role in the grounded theory approach. Case selection, or sampling, is not conducted based on a priori specifications (random or quota sampling, number of cases, etc.). Rather, it happens in the course of the research process (in parallel with data collection),

and corresponds substantively with the pursued or evolving theoretical considerations. The reasoning behind this is the idea of an iterative-recursive sequence of
research steps that can be viewed in analogy to the notion of a hermeneutic circle
movement of knowledge formation. An individual case is analyzed, and, in the
light of the ideas (interpretations, concepts, perspectives, etc.) generated in the
course of the analysis, it is decided and specified what empirical case(s) could
be interesting and enlightening as the next and subsequent cases for the development of a theory. This continues until a satisfactory degree of "saturation" of the
theory with a diversity of cases has been achieved.

However, in the approach practiced in the present chapter, the methodological
procedure remains rudimentary in this regard. The study presented here is merely
a single case analysis from which initial hypothetical deliberations on a theory of
the occupational-biographical development trajectories of memorial site guides
will be distilled. Compared with quantitative studies, in which mathematical-
statistical analysis procedures are used, sample sizes in grounded theory studies
are, as a rule, small. Nonetheless, they are always greater than one.

Despite this limitation, a first draft of a theory can still be developed based
on the analysis of a single case. However, if the development of the theory is
to be pursued further, this first draft must be followed by efforts to move the
research project forward in terms of case selection/sampling, thematic focus, category formation, and systematization. In that instance, it would be a matter of
considering what next cases could be selected for comparison. When doing so in
the present context, a survey methodology leap could be made into the field of
the activity/praxis of memorial site pedagogy in different contexts (places, coun-
tries/nationalities, institutional integration, guide characteristics, characteristics of
the visitors/groups, etc.). The type of data used for the grounding of a theory may
by all means change during the course of a research project.

The idea for the present study—to endeavor to take the first step in generating
a grounded theory based on the text of *The Memory Monster*—stems, first, from
the appeal of its extraordinary and profound account of work as a guide at Holocaust memorial sites, and, second, from my intuitive impression that conceptual
treasures lie buried within it that can be unearthed and made accessible for use
in developing a more far-reaching theory. In my treatment of the text, a certain
thematic focus is adopted: some things are emphasized, other things are left out.
Hence, a comprehensive appreciation of the text is not intended.

3 The Initially Self-Assured Guide—Academic Distance and Professionalism

"I burst into the field like a young bull" (Sarid 2020, p. 10).

When he first begins working as a guide, the personal stance profile displayed by the first-person narrator in Sarid's novella has the following briefly outlined characteristics, each of which is captured with a categorical keyword and a short excerpt from the English-language edition of the book (Sarid 2020).

The narrator's early search for an occupational field is informed by a superficial orientation aimed at personal well-being and taking life easy—easygoing stance.

"I envisioned myself sitting at a café in some tropical city in a light-colored suit, spending my days in elegant languidness, living on a modest yet respectable salary paid by the state" (Sarid 2020, p. 4).

His—in the light of the above motivation—unlikely decision to specialize in Holocaust studies as a source of income and a career domain is dictated by the necessity to provide for his family (he was engaged to be married)—job-like-any-other stance.

"I found myself specializing in Holocaust research out of practical considerations" (Sarid 2020, p. 4).
"The dean [...] said that meant I had one last realistic option for continuing my life as a historian in Israel—getting a PhD in Holocaust Studies. I was afraid" (Sarid 2020, p. 7).

The narrator's approach to topics or problems is characterized by a stance of academic distance. His Ph.D. focus is on the organization, technology, and logistics of the extermination work on the part of the perpetrators—academic focus of interest.

"I was drawn to the technical details of annihilation: the mechanism, the manpower, the method" (Sarid 2020, p. 8).
"[...] and then investigated the academic question of the variety that existed in methods of action; a surprising deviation from absolute unity, as one might have expected from an organization and a task of this nature" (Sarid 2020, p. 9).

The narrator's memorial-site-pedagogical concept is oriented toward the institutional regulations of the official mediation canon, which he refers to as a fixed script—conformist mediation stance: "I [was] holding strong to all of your messages, never deviating right or left. I was a good boy" (Sarid 2020, p. 15).

He pays hardly any pedagogical attention to the subjective characteristics of the high school students (and the other visitors) whom he encounters as a guide. He acts as a fact guide, and he expends little attention on the receptive willingness and processing capabilities of those he guides—focus on facts, not on teaching.

"[I] showered the children with my knowledge" (Sarid 2020, p. 10).
"I stood before them, trying to […] invade their hearts and their minds. I never felt like I truly succeeded, because I didn't love them enough. I know that now" (Sarid 2020, p. 15).

In his imagination, he is less empathetically mindful of the victims of the Holocaust, whom he encounters primarily as the many thousands, without differentiated individuality—defocusing the victims.

"Then the wing manager came to me [after a lecture to students at Yad Vashem]. She said you [the chairman of Yad Vashem] were impressed by my knowledge, but thought I lacked some emotion and personal attention to victims, *I'm a historian*, I thought, *not a social worker*, but I promised I would take that into consideration and try to correct my ways" (Sarid 2020, p. 12).

The protagonist did not choose the occupational field of Holocaust memorial work because of self-referential engagement—for example, with family history/generational connections (which inevitably existed because he is Jewish). He does not give any thought to his own involvement—or at least it does not feature in his self-presentation—bracketing of involvement: "I wanted to stay far away from disasters and calamities of our own people, guessing from the start the danger that lay in wait for me there" (Sarid 2020, p. 6).

The protagonist ignores—or dismisses as irrelevant—the psychological stress and the personal vulnerability that might result from constant intensive engagement with the events of the Holocaust—fiction of invulnerability: "I didn't take your warnings about the emotional strain [of working as a guide] too seriously because I had never suffered true emotional turmoil in my life, and thought I was immune" (Sarid 2020, p. 10).

At the beginning of the narrator's career as a guide, he presents the thematic/problem field addressed and his own person as quite distinct from each other—with no interpenetration or interconnection. Although he hints at a certain apprehension toward the Holocaust domain, he assumes that he has the matter under control, and that he can keep the existential explosiveness of the domain at bay by pragmatically contextualizing the guide job as a job like any other and by taking measures to ensure professionalism (through a conformist mediation stance), academic-scientific competence, distance, and interest focus. Feedback from others in his field of activity who point out the importance of soft factors (empathy, victim reference, own vulnerability) is ignored or rejected by him, and does not give rise to any self-reflection or self-doubt on his part.

I now seek an appropriate designation for a generalizing concept that is of a level of abstraction that fits the subject at hand, and under which the above-mentioned characteristics of Sarid's protagonist in his role as a guide can be subsumed. My first, and still provisional, attempt at a designation at this stage—what is this all about?—is personal (un)connectedness with (Holocaust) memorial work. In the present context, Holocaust memorial work is understood as a concept of personal action that includes, on the one hand, scientific engagement with the victim and perpetrator sides, with suffering and atrocities, with individuals and families, places and transports, roles and the division of labor, historical and political backgrounds, etc. On the other hand, it includes pedagogically conveying these things to visitors at the former concentration and extermination camps, preparing didactic-methodological procedures, designing memorial sites, and conveying the Holocaust in guide praxis. The term can also be developed further in the direction of—and can be linked to—societal (cultural, national, international) remembrance and memorial work.

Dimensionally, two poles of the concept Holocaust memorial work (in grounded theory terminology, a potential category) can be hypothetically differentiated: professional/academic distance und empathetic/identificatory closeness (Fig. 1). I metaphorically paraphrase the two poles of this dimension with the expressions keeping the thing at bay versus being involved with heart and soul. This dimension shall serve to characterize, on the one hand, the extent to which the narrator banishes his own emotionality—his own inner involvement in the horror scenarios of the Holocaust that are the subject of the job as guide—by using academic and pedagogical distancing procedures (methodological objectivation, focusing on "facts," generation of tunnel vision, etc.). On the other hand, it shall serve to characterize the extent to which the narrator incorporates empathetic,

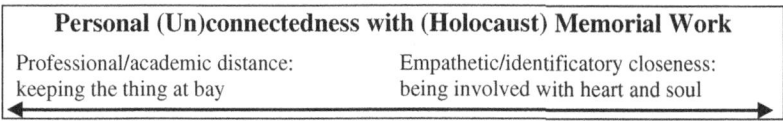

Personal (Un)connectedness with (Holocaust) Memorial Work

Professional/academic distance:	Empathetic/identificatory closeness:
keeping the thing at bay	being involved with heart and soul

Fig. 1 Differentiation of the concept of personal (un)connectedness with (Holocaust) memorial work in terms of a distance–closeness dimension

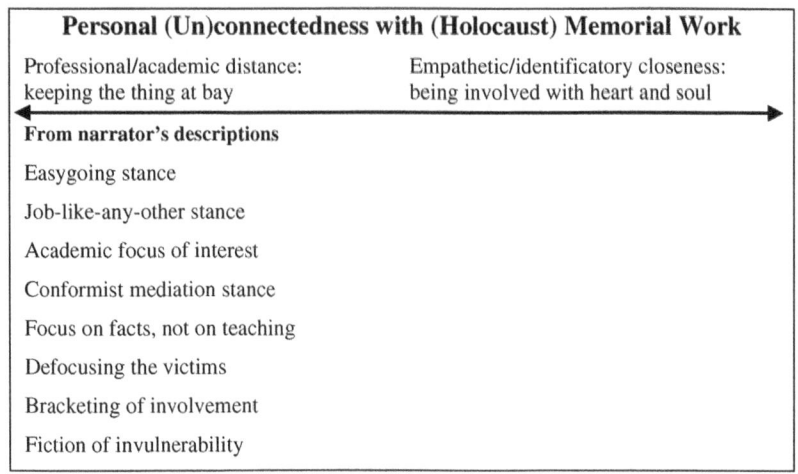

Personal (Un)connectedness with (Holocaust) Memorial Work

Professional/academic distance: Empathetic/identificatory closeness:
keeping the thing at bay   being involved with heart and soul

From narrator's descriptions

Easygoing stance

Job-like-any-other stance

Academic focus of interest

Conformist mediation stance

Focus on facts, not on teaching

Defocusing the victims

Bracketing of involvement

Fiction of invulnerability

Fig. 2 Differentiation of the concept of professional/academic distance on the basis of the descriptions provided by the first-person narrator in Yishai Sarid's novella *The Memory Monster* (2020)

compassionate emotional reactions (his own feelings of horror, compassion, grief, outrage, hatred, etc.) into his scientific and pedagogical activities.

The brief analytical sketch of Sarid's text in terms of the protagonist's initial and initiation period in research and pedagogical work in the Holocaust domain is summarized in Fig. 2 using the terminology distilled above, which should be seen as a preliminary stage in the development of an elaborated categorical vocabulary of a subject-specific substantive grounded theory.

During the initial phase in the field of activity as a Holocaust memorial site guide, Sarid's protagonist positions himself on the distance side of the distance–closeness dimension. The characteristics of closeness are a provisional construction on my part, following the heuristic grounded theory constant comparative maxim "search for the opposite." In terms of data, however, they have not yet been underpinned with empirical evidence. The differentiation of closeness remains hypothetical for the time being, and challenges me methodologically to engage in a thought-experimental way (on the basis of my relevant knowledge of the world, my field experience, etc.) in conceptual deliberations and to search for suitable, enriching, modifying, etc. examples in the data of our exemplary case (i.e., the protagonist's narrative) or in data from other empirical contexts (e.g., from data collections that we could undertake in the praxis field of memorial site work, from [literary] accounts of other contemporary witnesses, or from texts of scientific provenance). Thus, I follow the view that, for theory formation, the individual case, or its analysis, has the function of stimulating thoughts on conceptual structures that go beyond the reconstruction of the specific case, and that are also suitable for the presentation/explanation of other phenomena or case variants.

For the antagonistic manifestations of personal (un)connectedness—that is, distance and closeness—possible (sub)categorical or dimensional oppositional terms on the closeness side can now be elaborated. Fig. 3 outlines initial thoughts in this regard.

Personal (Un)connectedness with (Holocaust) Memorial Work

Professional/academic distance: keeping the thing at bay	Empathetic/identificatory closeness: being involved with heart and soul
◄───►	
From narrator's descriptions	**Hypothetical complement**
Easygoing stance	Stance with motivational references to the "weight of the world"
Job-like-any-other stance	Job with special challenge and responsibilty
Academic focus of interest	Focus on enlightenment and reconciliation
Conformist mediation stance	Mediation from an independent position and stance
Focus on facts, not on teaching	Empathy toward high school students, other visitors
Defocusing the victims	Victim references central to the job orientation
Bracketing of involvement	(Non)Identifying with victims or perpetrators, in relation to the background of own origin, family, generation etc.
Fiction of invulnerability	Awareness and mindfulness of own vulnerability

Fig. 3 Complementary, thought-experimental differentiation of the concept of empathetic/identificatory closeness

In Fig. 3, the characteristics of closeness are hypothetically constructed based on my own knowledge of the world and of the field. These characteristics must be empirically underpinned and differentiated. When doing so, it will likely turn out that among memorial site guides there are not only the two antagonistically pointed case types with the differentiated characteristics of distance and closeness but rather a variety of cases in between—that is, variants with intermediate forms of these characteristics, cases in which characteristics of one pole are combined or mixed with characteristics of the other, as well as cases that further expand the range of characteristics. There are also linkage possibilities here with the guide typology proposed by Ballis (2018).

Further questions can be raised that bring into focus the selection of next cases of interest, as well as thought-experimental, hypothetical explorations regarding empirical phenomena, relationships, or further conceptualizations. These questions include, for example:

- What are the (biographical, social, national, etc.) conditions and prerequisites for the occurrence of personal involvement, which is characterized by empathetic/identificatory closeness?
- What different process patterns in the occupational biographies of memorial site guides emerge when entrants characterized by professional/academic distance are compared with entrants characterized by empathetic/identificatory closeness?
- Can characteristic moments of endangerment of personal integrity (psychological stability, over-compensation, isolation, burnout, etc.) be identified for the two types when memorial site guides pursue this activity for a long time?

The question of how things continue with the protagonist's story and his initial self-assurance has already fed into the above deliberations. What becomes of his emphatically reiterated stance of distance toward Holocaust memorial work in the long run?

Whereas my reconstructions up to this point have gone mainly in the direction of a structural elucidation and modeling of the distance and closeness orientations of memorial site guides with regard to their personal (un)connectedness with thematic memorial work (a snapshot at the beginning of their careers, as it were), a new dimension is now added: personal development, change, transformation in this field of praxis and experience. The emerging theory of the stance and activity as a memorial site guide thus acquires an additional characteristic—namely, temporality and process.

4 The Memorial Site Guide in a State of Disorientation—Unsettling Moments

"The sports reporter, Slezak, stayed close to him [Bernays, an Auschwitz guide], although he did not ask any questions but only shook his head wordlessly, almost non-stop. After a while, Bernays, who had believed that he had long been prepared for anything, allowed this to irritate him after all. Can a person shake their head for hours without suffering any harm? That was a question that would surely have interested Mengele. When one was here for a long time, the associations simply did not manage to leave the reference space any more" (Menasse 2014, p. 79; translated from the German).

Sarid's protagonist's narrative reveals how his smooth facade develops cracks in the course of his work as a memorial site guide in Poland; how he is unsettled by certain experiences; how the once linear contours lose their clarity and definition; how he loses his seemingly fixed orientation; how the canonical business as usual of his pedagogical contact with his clientele is replaced with the unconventional and idiosyncratic; how his impatience and stridence grows; how he is in danger of floundering; and how—clinically speaking—he experiences loss of control and decompensation. The memory monster drives him on.

I use the categorical umbrella term unsettling moments to designate the phenomena that now become apparent. In the following, I will list the aspects through which, and the levels on which—according to the narrator's description—this dynamic is triggered und develops. Once again, I illustrate this with excerpts from the book. I provide an overview of the listed concepts (Fig. 4).

4.1 Experiences from Pedagogical Contact

The Israeli students' intrusively repetitive rituals of wrapping themselves in the national flag, singing the national anthem, and sometimes dancing with a Torah scroll at the memorial sites get on the narrator's nerves—Nationalistic and religious rituals.

"Wherever we went, they [the high school students] sang the anthem [the national anthem of Israel]. […] They spent most of their time in Poland cloaked in flags, singing. I talked to a teacher who had organized one of the delegations and asked her gently if we could cut down a bit [the frequency]. It sort of cheapens the anthem when you sing it two, three times a day, dozens of times a week" (Sarid 2020, p. 31).

The Israeli armed forces are planning to stage an incongruous memorial event in a former extermination camp in Poland to mark the occasion of the 75th

Unsettling Moments
Experiences from pedagogical contact
Nationalistic and religious rituals
Nationalistic staging of events
Chauvinistic slogans
Misdirected emotions of hatred
Transgressions
Moral deformation of the later-born generations
Self-experiences
Moral deformation in the private/family sphere
Self-suspicion of complicity
Fascination of the horror
Seduction by technological brilliance
Nightmares and hallucinations

Fig. 4 List of the unsettling moments of the memorial site guide

anniversary of the Wannsee Conference. The narrator considers the planned event to be a mockery of his idea of the character of the place and of the appropriate way of interacting with it—nationalistic staging of events.

The plan is to simulate a military operation where an Israeli army helicopter lands in Treblinka and soldiers pour out to rescue Jewish prisoners:

"The [press] officer took pictures of the memorial stones dedicated to exterminated communities. 'What do you think, should we build a set here?' she asked. 'We can build a few sheds for the soldiers to occupy; some guard towers, a bit of fencing? It's too empty this way. What do you think?'" (Sarid 2020, p. 138).

The narrator replies by urging her to listen for a moment to the sounds of nature, the wind and the birds, to transport herself back in time, to feel the presence of those Jews who had arrived at that very spot by train as subhumans, and who had ended up as worms, dust, and ground-up reptiles (Sarid 2020, p. 139).

The narrator overhears the gruesome chauvinistic platitudes and slogans, the craving for revenge and murder in the clandestine conversations of the high school students—chauvinistic slogans.

"For some reason, in Majdanek [...] I heard them talking about Arabs, wrapped in their flags and whispering, *The Arabs, that's what we should do to the Arabs*. Not always, not in all groups, but often enough for me to remember it. [...] Adults think the same things, but they keep it to themselves. Toward to the end [...] I gave my little speech outside the crematorium rather than join them inside. I didn't want to hear what they were saying in there" (Sarid 2020, p. 17 f.).

Among the high school students whom the narrator guides, the emotional side of the responsibility for the murders perpetrated in the camps—hatred for the perpetrators—is not directed at the Germans but rather diverted and misdirected in accordance with the students' own prejudice structures—misdirected emotions of hatred.

> "They didn't hate the Germans, the kids in my groups; not at all, not even close. The murderers barely registered in the narrative they [the students] created for themselves. They [...] never pointed an accusing finger at the perpetrators. They hated the Polish much more. When we walked around the streets in cities and villages, whenever we met the local population, they [the students] would mutter words of hatred at them, about the progroms they had committed, their collaboration, their anti-Semitism. But it's hard for us to hate people like the Germans. [...] We'll never forgive the Arabs for the way they look, with their stubble and their brown pants [...] The other part, intentionally and successfully planned by the Germans, was the fact that they committed their murder spree on Polish soil so they could keep Germany beautiful, clean, and well-organized. They kicked all the trash over to the east [...]" (Sarid 2020, p. 31 f.).

The high school teachers with whom the narrator deals during the guided tours pester him with know-it-all pedagogical advice, instructions, and other impositions—transgressions.

> "'I think you have to have more faith to the children,' she [a teacher] said. [...] 'Look into their eyes,' she proposed, and put her cold hand on mine in a gesture of mercy. 'That's what I do. That's how I connect to them'" (Sarid 2020, p. 62 f.).
> "They [these women] wanted me to comfort them after the difficult sights of the day, to explain to them how it was possible. Later, after we'd had a few drinks, they asked about my life, about my wife" (Sarid 2020, p. 19 f.).

When addressing questions such as "What do we learn from that time?" or "What is the historical lesson for the Jews and the State of Israel?" in his lectures to the students, the narrator experiences an unsettling victim-perpetrator identification between Jews and Nazis—moral deformation of the later-born generations.

On the last evening of each high school tour, a final discussion takes place with the high school students at the hotel. On one particular evening, the headmistress asks for a final comment:

> "A boy sitting on the sidelines stood up. [...] Somehow, I knew he would be saying something meaningful. 'I think that in order to survive we need to be a little bit Nazi, too,' he said. A bit of chaos ensued. Not too much, though. He was just saying to

adults what they [the students] usually only say among themselves. [...] 'What do you mean?' I asked. 'That we have to be able to kill mercilessly,' he said. 'We don't stand a chance if we're too soft.' [...] 'This is, after all, a war of survival. It's us or them. We won't let this happen to us again.' [...] 'Why the Nazis?' I asked. 'Why not the Americans, the Russians, the British? They were the ones that ended up winning the war.' The boy considered this. *Just don't chicken out on me now.* 'Because they went all the way,' he said" (Sarid 2020, p. 125 f.).

The monstrous and egregious résumé that Sarid's protagonist gives in this memorial-site-pedagogical situation is:

"Dear teachers, you can report back to your school that the message has been received. Only power. No conscience, no manners, no second-guessing. Those only challenge the soul and harm functionality. [...] We have to be a little bit Nazi. You've finally said it. You got the point, kids, well done" (Sarid 2020, p. 127).

4.2 Self-Experiences

The unsettling memory monster also infiltrates the narrator's private life—his family system. He goes into a rage when his wife tells him that their son, Ido, is being bullied and beaten by a group of boys at kindergarten. Inhibited against fighting back, Ido does not want to go to kindergarten any more. The narrator intervenes and menacingly threatens one of the bullies. Thus, his brutal stance toward the use of violence also permeates his private world—moral deformation in the private/family sphere.

"I returned to Israel urgently, having been summoned by Ruth. Ido was refusing to go back to kindergarten. Some boys were bullying him. [...] 'I'll go to kindergarten with you today and I'll take care of it,' I promised. [...] I asked Ido to the point out which kids had been hitting him. [...] I towered over that one boy, who finally looked scared, and shouted, 'Don't you dare touch my son!' His mother screamed, puffing up like a wild turkey, but I didn't care. The entire kindergarten railed around me. I didn't know these parents, but now they knew me. I stayed with Ido for a long time, until things calmed down and he agreed to say goodbye. Force is the only way to resist force, and one must be prepared to kill" (Sarid 2020, p. 65–67).

The question "What would I have done?" causes the protagonist great unease. He cannot absolve himself of the self-suspicion that, under those circumstances, he would have been prepared to keep quiet and participate—self-suspicion of complicity.

He arranges for the students to meet one of the Righteous Among the Nations, an elderly Polish woman, Anna R., at the farm where she still lives. At great risk to herself and her family, she had hidden a young Jewish boy in their barn for some months during the war. The teenager, a complete stranger, had managed to escape deportation and was on the run from the Nazis. Back at the hotel that evening, the narrator discusses Anna R.'s story with the students:

"'I ask myself [...] what would I have done in her place? I don't know. I would probably be too afraid to take the risk, and it's killing me, it won't let me go, because that's the only question we can ask ourselves as human beings.' [...] The more philosophical of the group would have rescued no one. Only the modest, the simple, the kind, would. I am not one of them, I told myself, and it made it difficult for me to carry on the conversation" (Sarid 2020, p. 59).

The narrator is alarmed to note that the accounts of Nazi atrocities—their sometimes perfect organization, their sophisticated logistics, and the decisiveness, relaxedness, and nonchalance with which the perpetrators implemented them—can exert an appeal on him, and that he sometimes experiences as sensual the description of these atrocities—fascination of the horror.

"Sometimes I got into the details of each step in the process, more than necessary, until the teacher or the commanders signaled to me that time was running out and that we had to move on. [...] For instance, I told them more than they needed to know about the haircutting process [...] How these stories riveted my twisted soul!" (Sarid 2020, p. 60 f.).
"And one last thing, which has slowly permeated me over the years, is the invisible admiration of the murder; the decisiveness and ruhtlessness, the audacity, the final, focused, and cruel act, after which there is nothing but silence" (Sarid 2020, p. 33).

The narrator is irritated by the ambivalence of his feelings when working as a consultant for an Israeli virtual reality (VR) company. The company is collaborating with the digitalization department at Yad Vashem on a project to develop what is ostensibly a computer simulation of an extermination camp for purely pedagogical purposes, but is in fact a computer game in which the players can assume the role of prisoner or guard. The first draft version of the game, which had already been developed before the narrator joined the project, was riddled with inaccuracies. He provides fundamental and detailed information about the conditions, processes, and division of labor in the extermination camps, thereby enabling the VR company to develop an improved version of the game. On the one hand, the narrator is fascinated by the technological brilliance of the simulation of the conditions and events in the camp, and admires the computer game for its closeness to reality, its detail, and functionalities. On the other hand, he is horrified by the possibilities for action and experience at the game console—seduction by technological brilliance.

This experience of ambivalence is related to his involvement in the project as an expert for the extermination process and to the seduction that the finished product exerts on him. The VR company sends him the revised version of the game based on his input:

"I tried it out. I played the part of a Jew, then of a German, and took some notes. [...] Jews could temporarily evade death through a few options in the operating software, for instance if they'd been chosen for hard labor or a medical experiment, or if they hid in a remote corner of the camp. The latter option did not exist in real life. I made an angry note to inform them to this. But all said, I was pleasantly surprised by their thoroughness: the game had all the components of the camp I'd described to them. [...] I pulled a gold tooth from the mouth of a corpse and placed it in a box. Then I switched to being German and whipped a Jew. Then I was a kapo [prisoner-supervisor] and ladled out soup. I couldn't stop—their game was so wonderfully terrible" (Sarid 2020, p. 115 f.).

The narrator is eventually afflicted by nightmares and hallucinatory visions in which the murdered victims from the extermination camps come alive, thereby acquiring a stature and voice of their own. During such episodes of the actualization of horror, the levels of reality become blurred; the protagonist reaches his psychological and physical breaking point—nightmares and hallucinations.

"The next day, in Auschwitz, I saw them for the first time. Not through books or the computer game, but for real. 'This is where the trains stopped,' I explained, hearing the train pulling up, the cars opening, seeing the floodlights, feeling the panic, *where's the kid, where's the suitcase, still alive. Where are we. Where do we go now.* I stood before my group and said nothing. I could feel their frantic movement around me. The explanations would wait. I was sick of the myth, the ideas, the perverted curiosity. I tried to hear what they were saying. [...] 'Grab him,' said one of the boys standing next to me. 'Catch him, he's falling.' I was gone for a few seconds, I'm not sure exactly how many. I woke up with my face wet. Above me was the strange sky. I tried to get up and the world shattered. 'Run to the entrance, have them call an ambulance, he's not well,' the doctor shouted above me" (Sarid 2020, p. 122 f.).

5 Action and Trajectories of the Work as a Holocaust Memorial Site Guide

"Bernays imagined that this was his last excursion. […] Auschwitz with him, that was a special experience; that was what everyone had assured him over the years: the students, the colleagues, the pensioners, the German, French, Jewish groups […] Maybe it was easy enough. To rebel, to desert, to break loose […]" (Menasse 2014, p. 89; translated from the German).

The activity as a memorial site guide extends far into the personal sphere—especially when the national, ethnic, or familial links between the guide and what happened during the Holocaust are as close as they are in the case of Sarid's protagonist. Despite his avoidance intentions and his initial belief that he is invulnerable, he learns this in a painful process. When he first started thinking about what profession he would choose, his deliberations were informed by an easygoing stance. However, the longer he works as a memorial site guide, the more his ethnicity and his past history catch up on him—in the form of a cascade of experiences of social and psychological irritation. The history of the Holocaust is an undeniable part of his identity, and this has become apparent to him in many ways. The memory monster has unexpectedly caught up with him. The unsettling experiences that he has in his job as a guide, and that radiate beyond the professional sphere, have affected his initial self-assurance and certainty and his stable action orientations.

Sometimes, he pro-actively confronts the unsettling moments that he experiences. In certain situations, he requests his clients to correct their behavior—for example, to respect the local circumstances. In addition, he revises his mediation strategies. The following are some examples:

- He tries to curb the excessive nationalistic and religious rituals of the visitors.
- He makes an effort to maintain his mental health balance: Sometimes he eavesdrops on the high school students' chauvinistic conversations.
- He pushes the topic of the use of violence as a survival strategy through pointed remarks and exaggerations.
- He no longer allows himself to be restricted by the provisions of the institutional memorial site guidelines.
- He radicalizes his forms of mediation: With time, he minces his words less and less (Sarid 2020, p. 59).

Increasingly, however, he experiences the fruitlessness of these efforts. He loses his firm hold, struggles to keep his balance, fights wildly against the unassailable, flounders mentally and physically. He observes in himself signs of disintegration:

"Talking began to weigh on me. Too many words. When I lectured, I listened to my own voice as if from the outside, like a person listening to themselves in a recording. It was grating" (Sarid 2020, p. 113).

He feels that, despite his best efforts, he is no longer able to connect with the high school students he guides:

"Sometimes I mustered up my strengh, shook myself awake, wore an amiable expression, searched for a way in, loosened my tongue. But they closed themselves to me, refusing to accept me. Their young faces looked like a minefield to me" (Sarid 2020, p. 114).

When he experiences failure or insult, he sometimes displays helpless, uncontrolled actionism:

"the manager [of the travel agency that hired him to guide the high school groups] told me that a few schools had sent bad reviews about me, saying there were some issues, and that they'd decided to put my services on hold for the time being. I was mad at them for doing this behind my back. I protested. No, sir, I wasn't going to be treated that way. I yelled at them. I lost my cool. I realized my reputation was ruined" (Sarid 2020, p. 142 f.).

He suffers a loss of moral integrity: own moral deformation, self-suspicion of complicity, fascination of the horror, seduction of technological brilliance. He experiences hallucinations by day and nightmares by night: the distressing appearance of the ghosts of the murder victims, which threatens his mental and physical health. And finally, he resorts to physical violence: the blows to the face of the German film director in Treblinka when he realizes the extent of the man's deception.

The proportion of the occurrences within the narrator's activity and action space that are subject to his voluntaristic control is progressively penetrated, overlaid, and limited by the proportion of occurrences that are driven by a logic of their own, that can to a large extent no longer be controlled by him, that are beyond his comprehension. He is increasingly at their mercy. Business as usual becomes increasingly impossible.

In the grounded theory literature, the process of being at the mercy of something, of quasi fatefully suffering something, has been conceptualized as a trajectory—that is, a process with a logic of its own that is (sometimes temporarily) split off from a biography, and over which the individual has no personal control. In such a process, a number of individual actions on the part of different actors may be interlinked without it being possible to identify a central subject. The protagonist is actively involved, but he or she has no control over the outcome. The concept was originally and primarily developed in the research of illness trajectories and processes of dying (Glaser and Strauss 1968; Corbin and Strauss 1988; Riemann and Schütze 1991; Schütze 1996; Strübing 2007, p. 117 f.). The notion of trajectories constitutes a theoretical bridge between the social science paradigms of structurally determined versus voluntaristic social action (Strübing 2007, p. 121). A characteristic pattern of development of trajectories can also be described (Schütze 1996, p. 129 f.).

Against the background of this notion of the relative proportions of (1) intentional action and action control and (2) determination by circumstances over which the individual has no control, a dynamic shift in this relationship can be described in the case of the narrator: his area of action control is restricted and reduced; the exercise of control by overwhelming circumstances, by these circumstances' own logic, increases. In Fig. 5, this notion is visualized in an overview sketch in which the relative proportions of the areas above and below the curve change over time.

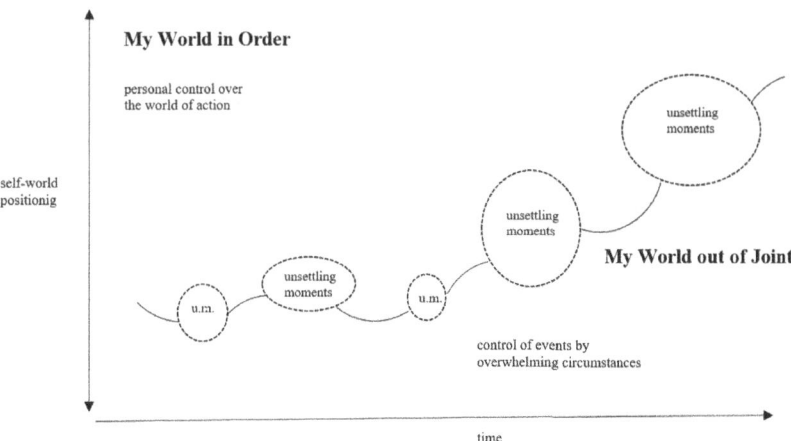

Fig. 5 Shifts in the relative proportions of action control and control by circumstances (u.m. = unsettling moments)

With regard to the development and transformation of the Sarid's protagonist, I distinguish two points in time: first, his entry into the occupation as a memorial site guide, described at the beginning of the narrative, when he charges into the arena like a young bull (Sarid 2020, p. 10); second, at the end of the narrative, when his professional situation has become precarious and questionable, and he is in an anxious, disoriented, and unsettled mental state. Accordingly, I identify two different world views, or conceptual self–world positionings, which—formulated from the narrator's viewpoint—I pointedly refer to here as "my world in order" and "my world out of joint." This fixation on two points in time is an idealization. The shape of the trajectory may differ; the steps or transitions in the curve may be flowing or abrupt, gradual or sudden.

The trajectory depicted in Fig. 5 shows mostly a steady rise, the speed of which increases over time and up to the end the narrative. This progression is intuitively sketched in this way in order to visualize the intellectual principle. The original shape of the curve is to some extent already marked out by the fate of the Jewish and other victims of Nazi German persecution who were excluded and torn from a world and a life as active members of society and increasingly subjected to state coercion, culminating in their deportation to and murder in the extermination camps.

If one uses such trajectory modeling to reflect on further cases (e.g., when looking for contrasting cases within the framework of theoretical sampling), different variants are conceivable. They include, for example:

• A sudden spike in the trajectory caused by a specific trigger situation: Sarid's protagonist recounts the story of a contemporary witness, a Holocaust survivor from Haifa, whom he invites to accompany a high school group to Auschwitz-Birkenau. On arrival at the extermination camp, the man is abruptly transported back in time to the scenes he experienced there as a child. From one minute to the next, he collapses.

• An oscillating movement, where periods of perceived personal control over the activity in this domain repeatedly give way to losses of control (crises, breakdowns), after which the protagonist finds his or her feet again, manages to continue on until the curve rises again, and so on. This trajectory is possibly characteristic of memorial site guides who pursue this activity for many years, and who from time to time experience a mental crisis during which they take time off or avail of supervision in order to be able to continue doing the job.

• Even a trajectory that leads from a pronounced "my world out of joint" state to a "my world in order" state and the (re)gaining of action control does not appear to be fundamentally impossible.

	My World in Order Personal control over the world of action	My World out of Joint Control of events by overwhelming circumstances
Personal unconnectedness with Holocaust memorial work: Professional/academic distance—keeping the thing at bay	**A**	**B**
Personal connectedness with the Holocaust memorial work: Empathetic/identificatory closeness—being involved with heart and soul	**C**	**D**

Fig. 6 Four-quadrant schema combining (un)connectedness with the concepts personal control and controlled by circumstances

Finally, the two components of the dimensional model sketched here can be combined for heuristic purposes (Fig. 6). The thought-experimental sorting of cases according to the logic of this four-quadrant diagram can be applied in order to reflect on known and familiar as well as not yet known or still unfamiliar cases, and to use this for further sampling deliberations (with a view to further empirical inquiry) and to gain new conceptual ideas.

Quadrants A and D of the schema are easy to fill when we think of the example of Sarid's protagonist:

- In the initial period of his activity as a memorial site guide, he lives in a state of personal unconnectedness in a world in order—quadrant A.
- After his personal involvement (connectedness) has become apparent, and the unsettling moments have become pervasive, he trundles into quadrant D.

But what about quadrants B and C? The case analysis so far has not yielded examples that would provide a concrete idea of the content of these two quadrants. But precisely for this reason, corresponding deliberations can be helpful in generating new concepts. The approach to developing a typology of memorial site guides proposed by Ballis (2018) can be incorporated here.

- Quadrant C could be home to the pedagogical ideal type of a memorial site guide: personally and empathetically connected with the problem field but at the same time capable of professional distance. One can conceive of a person who has a feeling for the victims and who displays an understanding of the students and other visitors, but who is also blessed with personal confidence and stability vis-à-vis the unsettling abyss of the material and the adversities of the mediation work.
- Quadrant B could accommodate, for example, a person who has been personally derailed by the subject of the Holocaust, and who—in an effort to escape this—tries to keep the matter at bay by using a very long tongs to self-distance him- or herself (denial, repression, etc.). In Eva Menasse's novel *Quasikristalle* (2014), we find the following sentence about her protagonist, the Auschwitz guide Bernays: "He himself knew exactly how one could cushion oneself against the horror with facts, figures, causal chains, and squadrons of footnotes" (Menasse 2014, p. 80).

6   Summary and Conclusion

In this chapter, I presented an initial theory sketch of the occupational-biographical development and transformation process of a professional Jewish-Israeli memorial site guide who leads tour groups through Nazi German extermination camps in Poland. I focused on the starting point, on the protagonist's stances and identifications, on the dynamics and the trajectory of the process, and on crises and ruptures and their causes.

Inspired by the phenomena described in *The Memory Monster* (Sarid 2020), I proposed theoretical concepts and process flows that can be conceived of as underlying the described phenomena. In the case analyzed here, a trajectory can be traced from a self-image of secure control over the professional activity, through an unsettling of perceptions, a shattering of the inner compass, and a loss of certainty of having intentional control over his own experience and action, to a state of being overwhelmed by circumstances and events outside his control. An empirical relationship can be hypothetically constructed with regard to the concepts of personal (un)connectedness with Holocaust memorial work and their respective manifestations closeness and distance.

The theoretical structures were developed here on the basis of a single case analysis. However, the grounded theory discovery methodology aims to go beyond the single case and to produce generalizable conceptualizations that are applicable to other cases and case trajectories. I believe that on the basis of

Sarid's text an occupational-biographical development type among memorial site guides can be characterized within the framework of the theory sketch presented here. I also believe that this type represents a variant in a spectrum of analogous types that differ, for example, in terms of their social, (sub)cultural, and national backgrounds, their trajectories in the respective proportions of personal control and control by circumstances, and further personal characteristics.

With regard to the demarcation of the scope of application of the present theory in progress, a number of deliberations are necessary. The demarcation will likely include a temporal-historical specification: the present theory sketch relates to the years at the turn of the century or the early twenty-first century.[1] The generational distance from those who were directly affected by the Holocaust plays a role, as does the frailty of the surviving contemporary witnesses who act as mediators of memorial site pedagogy, and the fact that their number is rapidly declining. A further factor is the temporal distance from the historical events, and the generational mentality of the memorial site visitors, which is shaped among others by the never-ending conflict with the Arab world. For memorial sites in other geographic locations, in other countries, and with visitors or visitor groups who have other configurations of characteristics, the reception stances will likely be specific to each site. Of interest is naturally the question of how such processes present themselves in a comparison of the (pre)histories of Jewish-Israeli guides with those of German guides (from different family backgrounds and with different generational distances): What consequences do these factors have for the trajectories of the activity as a memorial site guide (entries and exits, motivations, mission, mental balance, etc.)?

References

Ballis, A. (2018). Confronting subject matter education with memorial pedagogy. Guides at memorial sites and Holocaust museums. *RISTAL*, 1, 19–34.

Ballis, A. (2019). Guides an KZ-Gedenkstätten und Holocaust Museen—Professionalisierung in Zeiten eines Wandels der Erinnerungskultur. In A. Ballis, & M. Gloe (eds.), *Holocaust Education Revisited. Wahrnehmung und Vermittlung—Fiktion und Fakten—Medialität und Digitalität* (141–166). Wiesbaden: Springer.

Ballis, A., & Gloe, M. (2019). *Holocaust Education Revisited. Wahrnehmung und Vermittlung, Fiktion und Fakten, Medialität und Digitalität*. Wiesbaden: Springer.

---

[1] One temporal fixed point in the book is the preparation of the act of commemoration of the 75th anniversary of the Wannsee Conference, which took place in 1942. This points to the year 2017.

Breuer, F. (2009). *Vorgänger und Nachfolger. Weitergabe in institutionellen und persönlichen Bezügen.* Göttingen: Vandenhoeck & Ruprecht.

Breuer, F., Muckel, P., & Dieris, B. (2019). *Reflexive grounded theory. Eine Einführung für die Forschungspraxis.* Wiesbaden: Springer.

Charmaz, K. (2014). *Constructing grounded theory. A practical guide through qualitative analysis.* London: Sage.

Corbin, J.M., & Strauss, A.L. (1988). *Unending work and care. Managing chronic illness at home.* San Francisco: Jossey-Bass.

Dieris, B. (2009). *Sprechen und Schweigen. Aushandlungsstrategien des "Sich Kümmerns" um alte Familienmitglieder.* Hamburg: Kovac.

Glaser, B. (2005). All is data. *Grounded Theory Review.* http://groundedtheoryreview.com/2007/03/30/1194/. Accessed 1 April 2020.

Glaser, B.G., & Strauss, A.L. (1967). *The discovery of grounded theory. Strategies for qualitative research.* Chicago: Aldine Publishing Co.

Glaser, B., & Strauss, A.L. (1968). *Time for dying.* Chicago, IL: Aldine.

Menasse, E. (2014). *Quasikristalle.* München: btb.

Riemann, G., & Schütze, F. (1991). "Trajectory" as a basic theoretical concept for analyzing suffering and disorderly social processes. In D.R. Maines (ed.), *Social organization and social process: Essays in honor of Anselm Strauss* (333–357). New York: de Gruyter.

Sarid, Y. (2019). *Monster* (Ruth Achlama, German Translation). Zürich: Kein & Aber.

Sarid, Y. (2020). *The Memory Monster. A Novel* (Yardenne Greenspan, English Translation). Brooklyn: Restless Books.

Schütze, F. (1996). Verlaufskurven des Erleidens als Forschungsgegenstand der interpretativen Soziologie. In H.-H. Krüger., & W. Marotzki (eds.), *Erziehungswissenschaftliche Biographieforschung* (116–157). Opladen: Leske und Budrich.

Strauss, A.L. (1987). *Qualitative analysis for social scientists.* New York: Cambridge University Press.

Strübing, J. (2007). *Anselm Strauss.* Konstanz: UVK.

North America & South Africa

Self-guiding, Moderating, Accompanying, and Survivor-Guiding

Educational Approaches at the Sarah and Chaim Neuberger Holocaust Education Centre in Toronto

Franziska E. Müller

Abstract

This article will discuss the different approaches to guiding at the Sarah and Chaim Neuberger Holocaust Education Centre in Toronto: Self-guiding, moderating, and accompanying. Since the Neuberger is not a large museum with numerous guides, the pedagogical program of the Neuberger will also partly be considered—this includes a discussion of the Historical Thinking Concept (HTC) and one of the Neuberger's digital educational resources. Lastly, this paper will take into account yet another way of guiding—survivor guiding. In this context, the focus will be on Pinchas Gutter who lives in Toronto and often speaks at the Neuberger. This notion of guidance by a survivor will be correlated to the work of guides at the Neuberger. This article is therefore similar to a case study, based on the three interviews. These were analyzed content-based to provide insight into the educational approaches at the Neuberger.

1 The Sarah and Chaim Neuberger Holocaust Education Centre

Thursday, September 7, 2017: The sounds of a harmonica playing a cheerful tune fill the foyer of the Sarah and Chaim Neuberger Holocaust Education Centre in Toronto (hereinafter Neuberger/Centre); George Scott, an 87-year-old survivor of

F. E. Müller (✉)
LLM student at European University Viadrina, Frankfurt an der Oder, Germany
e-mail: franzi-mueller.12@gmx.de

© The Author(s), under exclusive license to Springer Fachmedien Wiesbaden   179
Gmbh, part of Springer Nature 2022
A. Ballis (ed.), *Tour Guides at Memorial Sites and Holocaust Museums*,
Holocaust Education – Historisches Lernen – Menschenrechtsbildung,
https://doi.org/10.1007/978-3-658-35818-1_10

the Holocaust, demonstrates his musical talent. He and his daughter took the time
to meet my family and me at the Neuberger.[1] At the time 19 years old, I had
finished my voluntary year of social service at the Memorial Site Dachau only
a few days before. As part of a project focusing on the history of children and
young adults in the concentration camp Dachau, I had spent months researching
what had happened to George during the Holocaust. The Managing Director of
the Neuberger, Dr. Carson Phillips, had helped me a great deal with my research
and also connected me with Mr. Scott. Over the years, Philipps and I remained in
contact and it is therefore my pleasure to contribute an article about the Neuberger
and its work. For this article, I conducted interviews with him, Holocaust survivor
Pinchas Gutter, and Audrey Diamant, a guide at the Neuberger.

The Neuberger is located in an outer district of Toronto where many residents are members of the Jewish community (Interview with Phillips, 30 July
2020, 0:50–1:00). It was opened in 1985 and features a museum and a library.
The Centre offers a wide range of educational programs such as guided tours
through the museum, workshops, and digital resources. Moreover, the Neuberger
is especially known for what it refers to as its "signature program"—the annual
Holocaust Education Week (HEW). HEW includes a broad variety of events such
as lectures, films screenings, discussions with survivors, exhibits, and book talks
to learn about the history of the Holocaust and to discuss Holocaust education
(UJA Federation of Greater Toronto & Sarah and Chaim Neuberger Holocaust
Education Centre 2019). Especially school groups visit the Neuberger (Interview
with Phillips 30 July 2020, 1:00–1:30)—an estimated 20,000 students each year
(www.holocaustcentre.com/about-us/who-we-are). School groups typically do a
guided tour at the museum, which may also include talking to a survivor (www.
holocaustcentre.com/educators-students/field-trip).

This article will discuss the different approaches to guiding at the Neuberger:
Self-guiding, moderating, and accompanying. Since the Neuberger is not a large
museum with numerous guides, the pedagogical program of the Neuberger will
also partly be considered—this includes a discussion of the Historical Thinking
Concept (HTC) and one of the Neuberger's digital educational resources. Lastly,
this paper will take into account yet another way of guiding—survivor guiding. In

---

[1] Born in Budapest in 1930, George Scott was deported to Auschwitz in 1944 and then
forced to work in the subcamps of the Dachau concentration camp. On the 29th of April
1945, the US Army liberated the Dachau concentration camp. George went back to Hun-
gary, then stayed in Germany for a while and eventually emigrated as an orphan to
Canada in 1948. After retiring he began speaking at the Neuberger Centre (Montréal Holocaust Museum 2017) http://holocaustlifestories.ca/wp-content/uploads/2017/08/Biography-
George-Scott.pdf.

this context, the focus will be on Pinchas Gutter who lives in Toronto and often speaks at the Neuberger (Phillips, personal communication 6 September 2020; Körte-Braun 2013). Gutter's testimony was the first to be recorded in the form of an "interactive biograph[y]" (USC Shoah Foundation 2018). His testimony laid the foundation for what is now the USC Shoah Foundation's *Dimensions in Testimony* project (de Jong 2018, p. 247; Körte-Braun 2013; Ballis in this volume). In 2017, Pinchas Gutter took part in the creation of yet another pioneering means to preserve his testimony: The 20-min virtual reality film *The Last Goodbye* enables viewers to accompany Pinchas as he walks through what was once the Majdanek extermination camp (de Jong 2018, p. 248; Watercutter 2017). This notion of guidance by a survivor will be correlated to the work of guides at the Neuberger and hence demonstrates the multifaceted ways of guiding. This article is therefore similar to a case study, based on the three interviews. These were analyzed content-based to provide insight into the educational approaches at the Neuberger.

2   Guides at the Neuberger

Before the SARS-CoV-2-pandemic, an estimated two or three groups visited the Neuberger each day of the week—usually school groups (Phillips, personal communication 20 August 2020). Phillips describes the Neuberger:

> "[It's] not really like a drop in-Centre where visitors might just decide to pop in [...]. So the Centre, when it was first developed, which was back in 1985, was developed as a place where groups would come—school-groups primarily was the target audience—and it has remained [...] pretty much in that way [...]" (Interview with Phillips 30 July 2020, 0:35–1:16).

Most school groups are grade ten students where the history of the Holocaust is a mandatory part of Ontario´s curriculum (Interview with Phillips 30 July 2020, 1:50–2:00, 26:50–27:05; *The Ontario Curriculum, Grades 9 and 10: Canadian and World Studies, Geography, History, Civics (Politics)*: 119, 128, 141, 147). School groups typically do a tour at the Neuberger which may vary in its duration but is usually scheduled for two-and-a-half hours. A tour includes a brief introduction to the history of the Holocaust and the thematic focus of the tour, followed by a six-minute film that summarizes the history of the Holocaust. The group will explore the exhibition and discuss maps and statistics—for example, a map displaying the number of Jewish victims of the Holocaust per country. Afterwards, students may talk to a survivor, which usually takes about 45 to 60 min (Interview with Diamant 7 August 2020, 21:00–26:15; Interview with Phillips 30 July 2020, 4:00–5:30). The tour specifically discusses the experiences of Jewish people during World War II, yet other victim groups are also briefly touched upon (Interview with Diamant 7 August 2020, 21:35–22:10). The introductory film is supposed to be the foundation for a discussion among the students (Interview with Diamant 7 August 2020, 22:10–22:38; Interview with Phillips 30 July 2020, 4:00–5:15).

Guides at the Neuberger are non-employed volunteers, "most of [...] [whom] have a family connection to the Holocaust—they're either the children or the grandchildren of Holocaust survivors, [...] so they come with [...] [a] personal interest" (Interview with Phillips 30 July 2020, 5:48–6:00, 6:04–6:31) to educate students on the history of the Holocaust. Currently, there are 16 volunteers at the Centre. Many of the guides are retired and around 50 to 65 years old—they have been conducting tours at the Neuberger for a long time, resulting in the Neuberger having a relatively consistent group of guides. Many of the volunteers have close ties to the Jewish community (Interview with Phillips 30 July 30, 2020, 6:50–7:35, 22:25–22:35; Phillips, personal communication 6 January 2021). The family connection of the guides to the topic also influences how they interact with students—it allows the guides to approach the history of the Holocaust from an individual perspective: Historical facts are combined with the individual experiences of their relatives, and as a result, their perspective is not a purely historiographic one. Audrey Diamant, who has worked as a tax partner at PwC and has been a volunteer at the Neuberger for 16 years now (Diamant, personal communication 4 January 2021), says that the impact of her parents' past—who were both Holocaust survivors—on her work at the Centre is a "moment of [...] passion in terms of how I speak because it's through my parents at least a bit more of a lived experience" (Interview with Diamant 7 August 2020, 0:14–0:20, 9:35–9:53). Audrey's mother had survived Auschwitz and other concentration camps, and her father was forced to work as a slave laborer (Diamant, personal communication 20 and 23 January 2021). Both were originally from Romania where they—after the war had ended—had been living again until 1948. After an eight-month stay in Israel, they emigrated to Canada in 1950 (Diamant, personal communication 4 January 2021).

However, she mentions her parents only occasionally "depending on the circumstances [...]—if I think it's relevant to relate it [the history] a little bit to what I heard from my mother or my mother-in-law [...] I might make brief mention" (Interview with Diamant 7 August 2020, 15:40–16:55). Sharing personal information with the group adds a notion of reality, of actuality to the purely historic facts—what the guide tells becomes more tangible for students (Interview with Diamant 7 August 2020, 17:10–17:28).

Notably, the guides not only interact with visitors, they also act as a bridge between the Neuberger and the Holocaust survivors (Phillips, personal communication 20 August 2020), as Audrey explains:

> "One of the programs we have, [...] which is just a wonderful idea, is that every museum educator is kind of responsible if you will for three or so survivors. And what that means is that once a month you call them 'How are you? What's new?' [...] we also report that so the Centre knows that they're okay. But in the course of that you also kind of develop friendships and there are a number of them that I'm quite close to. One of them is 95 years old and she, she used to go to spinning classes with me [...] the 95-year old with the little tattoo on her left arm [...] is in my spinning class! And when I'm getting tired I look over at her and she keeps going and so she humiliates everybody [...] she's amazing!" (Interview with Diamant 7 August 2020, 35:55–36:48)

The volunteers are not only guides at the Neuberger but also contact for and even friends with the Holocaust survivors—being a volunteer at the Centre also means new extracurricular activities. Survivors, guides, and the Neuberger staff share very friendly and close relations with each other—they are a closely-linked team.

The structure of the guided tour highlights the multiple functions of the guides even more since guiding at the Neuberger draws on various different approaches as well. Probably the most common way to guide a group is by means of telling visitors about the history of the Holocaust in a reciting manner—in colloquial terms, the guide acts as the "sage on the stage" (King 1993, p. 30). In educational sciences this technique of teaching or, in this case, guiding, is called the "transmittal model"—"the one who *has* the knowledge [...] transmits that knowledge to the students" (King 1993, p. 30). Of course, this method is applied at the Neuberger as well, since the transmittal model is especially convenient for introducing students to the topic and ensuring that everyone has the same level of knowledge (Interview with Diamant 7 August 2020, 21:20–21:35).

Yet another approach to guiding is to let students guide themselves: Smaller groups of students become experts for a specific exhibit and will have to share their takeaways with the entire group once they get back together. Students will get information from the exhibit description, discuss the exhibit in their group and later present it before the others. Especially during this part of the tour, the guide tries to get students to participate in the discussion and moderates the debate (Interview with Diamant August 7, 2020, 25:05–26:00). Other than the transmittal approach, this method, which is called the "constructivist model," relies on the "students' interaction with the material and with each other" (King 1993, p. 30).

The guide stays on the sidelines and might answer comprehension questions, but in essence, the students are supposed to guide themselves—eventually they convey to their fellow students what they were able to learn about a particular exhibit. Hence, throughout the tour, the task of the guide shifts from being the "sage on the stage" to being the "guide on the side" (King 1993, p. 30), who moderates the discussion and gives assistance to the students.

Apart from historic facts, the tour and also of many of the Neuberger's programs focus on discussing the choices people had during the Holocaust (Interview with Diamant 7 August 2020, 59:00–1:01:47; Interview with Phillips July 30, 2020, 48:00–51:45). Audrey Diamant explains that she wants students to think about what they would have done, about the choices they would have made, had they been living at the time of the Holocaust:

> "It's, I think, all part of […] getting them to think a little bit about the experience itself, what might have happened, and how they would react to it as opposed to just lecturing to them. And, and a lot of times, you know, you ask them bigger questions as in 'What would you do?' You know, give it some thought, […] 'Would you be a bystander, or onlooker?' […] 'Honestly, how do you feel and why?' And then you kind of engage them" (Interview with Diament 7 August 2020, 1:01:15–1:01:49).

Since the Neuberger museum is neither located in a country where National Socialism was able to prevail for twelve years nor at a place with a specific historic past in regard to the Holocaust, a guided tour at the Neuberger is already based on a certain distance to the topic. Therefore, this approach of asking students to consider how they would have felt and what they would have done allows them to engage more with the topic and overcome a potential distance to the topic created by time and space.

The Neuberger refers to the volunteers who conduct the tours as "museum educator[s]" (Interview with Phillips 30 July 2020, 4:10–4:18) rather than guides. Reflecting on the structure and content of the tour raises the question how the museum educators are instructed. Interestingly, educators at the Neuberger do not experience a standardized training to conduct tours:

> "We have designed, I would say, […] it's not really a formal training program, it's more of an informal training program, where we work with them based upon what their knowledge level is and supplement that with additional historical information. […] We've done this over the years in sort of a group setting to […] establish a […] historiography that is consistent amongst all the guides. We work with a lot of the writings of Doris Bergen—[…] she wrote a book called […] *War and Genocide. A Concise History of the Holocaust*, […] it's a very good introductory text for […] a

lot of educational purposes. So we use that as sort of the foundational piece for having all of the guides on the same page when it comes to discussing the history of the Holocaust" (Interview with Phillips 30 July 2020, 6:30–9:08).

Bergen is a professor at the Department of History, University of Toronto (Bergen 2016, p. 361). Her book is indeed very approachable, regardless of the preexisting knowledge level about the history of the Holocaust—this is also due to the fact that her book "takes an integrated approach, drawing on personal accounts to not only illustrate but also complicate general claims" (Bergen 2016, p. 1). Her book is a good resource for the museum educators, and "undergraduate courses in North America use it as a textbook" (Interview with Phillips 30 July 2020, 8:20–8:30) as well.

Because of their personal connection to the history of the Holocaust and the contact to the survivors who speak at the Neuberger, guides already have considerable knowledge about the topic. Audrey Diamant explains:

"I had always been, say, interested in the Holocaust, one might argue obsessed by the Holocaust, since I was a, a little girl. My parents were Holocaust survivors, [...] I lost a lot of family in the Holocaust [...] and I saw this [working as a museum educator at the Neuberger] as a way to do something a little bit more concrete about this area of interest. I thought it would help me improve my knowledge but also then do something concrete in terms of educating or helping to educate children" (Interview with Diamant 7 August 2020, 0:00–0:47).

When Audrey became a guide at the Neuberger, she learned how to do tours by accompanying other guides who had been at the Centre for a long time until she eventually began to do tours herself. Every so often, the guides will meet with the Neuberger's staff to do an "educational session[s]" (Interview with Diamant 7 August 2020, 1:55–3:10, 57:10–57:35) where they discuss new material or specific topics. In addition to the historical knowledge, the guides are also introduced to the pedagogical techniques of guiding, more specifically how they can interact with students and what questions they could pose in order to engage them (Interview with Diamant 7 August 2020, 1:27–2:11, 55:30–56:26). Audrey Diamant explains that educating students about the history of the Holocaust is a personal concern of hers and also a responsibility to educate herself as much as possible in order to be able to answer the students' questions (Interview with Diamant 7 August 2020, 0:00–1:00, 3:20–6:55). All in all, the training of guides at the Neuberger adapts to the individual pre-existing knowledge of the guides and adds additional historiographic and pedagogical knowledge. The transfer of knowledge to the guides at the Neuberger is combined with the guides' motivation to educate themselves out of their personal interest.

Apart from these training-methods, the Neuberger currently uses the *Historical Thinking Concepts* (HTCs), as "one of the guiding principles for [...] train[ing] the guides" (Interview with Phillips 30 July 2020, 19:00–19:08).

3   The Historical Thinking Concepts

In recent years, the HTCs became part of the Ontario-curriculum—grade ten history now includes the HTCs (*The Ontario Curriculum, Grades 9 and 10: Canadian and World Studies, Geography, History, Civics (Politics)* 2018, 103–106). The aim of the HTCs is for students to learn history through the use of primary sources (Interview with Phillips 30 July 2020, 9:00–10:23).

The concepts—which are meant to improve the methods of teaching history in Canada—are the result of the *Historical Thinking Project* (2006–2014), led by Peter Seixas of the University of British Columbia (https://historicalthinking.ca/about-historical-thinking-project). In summary:

"[The] Historical Thinking Project works with six distinct but closely interrelated historical thinking concepts. To think historically, students need to be able to:
1. Establish historical significance
2. Use primary source evidence
3. Identify continuity and change
4. Analyze cause and consequence
5. Take historical perspectives, and
6. Understand the ethical dimension of historical interpretations"

(http://historicalthinking.ca/historical-thinking-concepts).

The HTC relies on the use of primary sources in class—for this purpose, the Neuberger Centre developed the *Brady Resource Kit* so that classes can work with replicas of primary sources, which in this case are mostly documents and photos (Interview with Phillips 30 July 2020, 11:25–11:48). George Brady was a Czech Holocaust survivor and the brother of Hana Brady who became widely known in recent times since the protagonist of the book *Hana's Suitcase* is based on her (Interview with Phillips 30 July 2020, 11:40–12:10). *Hana's Suitcase* was written by Karen Levine and published in 2002—the popular book was later translated into many other languages and became the model for a documentary film and a theater play among others (Sarah and Chaim Neuberger Holocaust Education Centre n.d., p. 4, https://www.hanassuitcase.ca/).

George and Hana were deported to the Terezín camp in 1942—both were sent to Auschwitz in 1944 where Hana was murdered. George survived the Holocaust and—only 17 years old at the end of the Second World War—eventually emigrated to Canada in 1951 (Sarah and Chaim Neuberger Holocaust Education Centre n.d., p. 4; https://www.hanassuitcase.ca/).

Due to the fact that the Brady family and their neighbors saved a lot of documents from the time of the Second World War, the *Brady Resource Kit* is able to offer a wide range of digitized versions of them (Interview with Phillips 30 July 2020, 13:55–14:40). The online platform offers four thematic envelopes covering the years before 1939, the years between 1939 and 1941, 1942 to 1945 and the postwar years. In addition, educators receive a teacher guide and a PowerPoint presentation to work with in class (www.bradyresourcekit.ca/geo rge-brady-kit). Each envelope contains digitized documents or photographs of objects—by clicking on them one is provided with a brief document description or also a brief video clip where George Brady talks about the document himself (https://www.bradyresourcekit.ca/object-4-8,https://www.bradyresourcekit.ca/object-3-9,https://www.bradyresourcekit.ca/object-2-1). Since many documents are either in German or Czech, students would have to translate them:

"We have them think through the process: 'Ok, does anyone else in your class perhaps speak Czech? Or does anyone have a grandmother or a grandparent who speaks Czech?' […] Because Toronto is such a diverse city, sometimes you will have someone in that class who can speak Czech […]. Or, […] some of them will say, 'I have a translation app on my phone, let me see if I can do it.' […] It's more so the process of getting them to think it out than it is to give them all the answers upfront" (Interview with Phillips 30 July 2020, 17:50–18:35).

Discussing an example of how students work with the Brady Resource Kit shows the implementation of the HTC: One of the digitized documents is a letter denouncing the Jewish Brady family. The letter had been written by one of the neighbors and was intercepted by the postmaster before the Nazis were able to receive it—in fact, the postmaster not only kept the letter, he also warned George Brady's father that his family was in immediate danger of being betrayed to the Nazis (Interview with Phillips 30 July 2020, 48:35–50:46, https://www.bradyreso urcekit.ca/object-2-1). Based on the HTC, students would see the digitized version of the document and translate it. Since this is a handwritten letter, which poses another difficulty in addition to the language barrier, they might receive a little more help than they would with other documents. Once they have a translation of the letter, they would discuss the origins of the document and also the choices of the people involved in this incident: What motives did the neighbor pursue when he wrote the letter and brought it to the post office? Why did the postmaster decide to give the letter to the Bradys instead of processing the letter in accordance with the official regulations (Interview with Phillips 30 July 2020, 50:44–51:21)? The Neuberger uses this particular document to show that "people have choices along the way—they're not completely helpless and all of these choices have consequences" (Interview with Phillips 30 July 2020, 51:21–51:39). Therefore, most of the six historical thinking concepts are directly applicable— by using a primary document, students discuss the "cause[s] and consequence[s]" (http://historicalthinking.ca/cause-and-consequence) of the neighbor's action and see the incident from a "historical perspective." Moreover, the students consider the "ethical dimensions" of the occurrence by asking what this means for today's society and for values such as courage and solidarity. By acquiring knowledge about the Holocaust on their own, the learning effect will last longer (King 1993, p. 30, http://historicalthinking.ca/historical-perspectives, http://historicalthinking. ca/ethical-dimensions).

While working with the HTC is beneficial for students, it also means a change for the guides regarding their work:

"[It] was a challenge for some of our guides, especially the ones who had been there longer because they want to tell the student what is so important, what is significant about this item but that is not the approach of historical thinking. Historical thinking says you let the student uncover what is important about it and teach them ways of how they're going to find the information. [...] It's not always easy because I think a lot of us when we have this information, we want to share the information and what we have to do is sort of stand back and help the students to come to the conclusions on their own" (Interview with Phillips 30 July 2020, 17:15–19:27).

This leads back to the principles discussed before—implementing the HTCs relies on guides being the "guide on the side" instead of the "sage on the stage" (King 1993, p. 30).

4   The Survivor Guide? Pinchas Gutter and Digital Guiding

At Holocaust memorial sites it is not uncommon for Holocaust survivors to accompany visitors through the memorial site. The Neuberger provided different means to visit a concentration camp memorial site by offering a screening

of *The Last Goodbye* as a pilot project for a month (Phillips, personal communication 20 August 2020). Before discussing the virtual reality film, I will give a
brief overview of Pinchas' biography.

Pinchas Gutter and his twin-sister Sabina were born on July 21, 1932 into a
Hasidic family, resident in Łódź, Poland (Ballis 2021, p. 4; Gutter 2017, p. 24).
When the Second World War broke out, the Gutters moved to Warsaw where they
had relatives. In November 1940, the Nazis set up the Warsaw Ghetto, and the
Gutter family was forced to live there until they were deported to Majdanek in
May 1943, after the Warsaw ghetto uprising. There, Pinchas' sister and his parents
were murdered immediately after they had arrived. Pinchas was a few months in
Majdanek before he was brought to the forced-labor camps Skarżysko-Kamienna
and Częstochowa (Ballis 2021, p. 4 f.; Gutter 2017, p. 68–70, 72, 74–77, 91; UJA
Federation of Greater Toronto et al. 2018). When the Soviets gained ground, the
captives of the Częstochowa camp were brought to the Buchenwald concentration
camp and from there to the Colditz concentration camp. In April 1945, as the
war was coming to its end, Pinchas was forced on a death march from Colditz to
Terezín, where he was liberated by the Soviet army soon after they arrived (Ballis
2021, p. 5; Gutter 2017, p. 94–97, 100–101; UJA Federation of Greater Toronto
et al. 2018). After the liberation, Pinchas was among those orphans who were
sent to Britain under the auspices of the *United Nations Relief and Rehabilitation
Administration*—after staying in Britain for some time he left for France where
relatives of his were living. He spent some years living in Israel, Brazil, and
South Africa before he and his wife Dorothy eventually emigrated to Canada in
1985 (Ballis 2021, p. 5 f.; Gutter 2017, p. 103, 109–111, 115–117, 122 f.; UJA
Federation of Greater Toronto et al. 2018).

In 1967, he spoke about his experiences before a larger audience in East London for the first time—he had been asked to speak on Yom HaShoah. The event
was such a traumatic experience for him—he suffered from "horrible nightmares
and flashbacks" (Gutter 2017, p. 119)—that he did not speak again about his life
in public until the beginning of the 1990s. Nevertheless, he managed to deal with
the trauma, and after he had moved to Canada, he began speaking about his experiences—interviewed by historian Paula Draper in Toronto, his first interview took
place in 1993, and he eventually became a speaker at the Neuberger Centre (Ballis 2021, p. 6; Interview with Gutter 11 September 2020, 0:00–2:41, 7:39–8:40;
Gutter 2017, p. 120–122). In 1998, he was introduced to Stephen Smith who was
involved in establishing a Holocaust museum in Cape Town at the time—since
that time, they have collaborated on many projects (Interview with Gutter 11
September 2020, 30:55–31:55). The interactive biography and the virtual reality
film are the most widely known of these projects—in both cases Pinchas was the

first Holocaust survivor to participate in the undertaking (Interview with Gutter 11 September 2020, 19:45–21:15, 28:05–28:20).

Through the use of 3-D-scans and about 30,000 images of Majdanek, the virtual reality film enables viewers to walk next to Pinchas through the former camp by putting on a virtual reality headset (Watercutter 2017). The filming at the Majdanek memorial site took three days—as Pinchas recalls he "had to suffer three days by feeling very alone and reliving, almost like arriving in Majdanek, [...] in the cattle car and with dead people inside." However, he "knew that [...] in the future it's going to be an educational tool and for [...] [him] that is the most important aspect of it" (Interview with Gutter 11 September 2020: 37:13–37:27, 37:00–37:10)—in other words, the educational value of this project outweighed the personal pain it caused Pinchas. For Pinchas, filming at the memorial site and talking to the cameras as if people were listening was entirely different as opposed to being at the memorial site with an actual group (Interview with Gutter 11 September 2020, 1:08:25–1:09:25). Pinchas sometimes accompanies groups to memorial sites—he is a participant of the *March of the Living* and in 2016, for example, he went to Poland with the *Study Tour for Holocaust Educators* (Rubenstein & March of the Living 2020, p. 64 f.; Phillips, personal communication 6 September 2020, March of the Living Holocaust Survivors Database 1988–2019. (n.d.)). The Neuberger offers the *Study Tour for Holocaust Educators* as a "unique educational program that brings teachers, teacher candidates, and faculty to sites of Jewish memory in Austria and Poland" (https://www.holocaust centre.com/educator-development). Travelling with such a group and visiting the Majdanek memorial site with them is a much more personal experience; Pinchas explains that "you try to get them to identify themselves with you" (Interview with Gutter 11 September 2020, 1:05:40–1:05:50). In other words, during the travelling and the visit, a connection between Pinchas and the groups develops—this interaction as well as Pinchas' sharing details about his life with the group make the tour a joint experience. Moreover, both Pinchas and the participants of the study tour profit from their mutual experiences since he says that "the empathy that I got from them helped me healing [...] and the knowledge that I imparted to them helped them educate people from hearing it from a [...] witness" (Interview with Gutter 11 September 2020, 6:00–6:16). However, filming *The Last Goodbye* left him "feeling very alone" (Interview with Gutter 11 September 2020, 37:16–37:20), as there was no interaction with a group, no joint experience. For Pinchas, filming was an individual experience as is watching the film for the viewer—for him, the direct emotional connection between him and the participants was missing (Interview with Gutter 11 September 2020, 1:08:40–1:09:24). In contrast, viewers do develop such a connection to some degree since the film creates the illusion that Pinchas is actually there to accompany them through what was once the Majdanek camp—an effect that is called "emotionale[n] Immersion (emotional immersion)." Rothstein notes that the virtual reality film turns viewers into witnesses of the survivor's experience (Rothstein 2020, p. 217).

Keeping in mind the different methods of guiding, the virtual reality film *The Last Goodbye* offers yet another approach. Watching the film, one is under the impression that Pinchas is guiding the viewer through the memorial site. While I would call this approach *"survivor-guiding,"* Pinchas pointed out that he does not see himself as a guide in the way guides at memorial sites or museums are and that there is an important distinction to make:

> "I think […] that the role of the Holocaust survivor is not to be a guide—there were […] guides there that were talking about generalities about what happened in the different places; but what you try to do, at least what I try to do, is to actually marry the person to the place. So, in other words when you go to Majdanek […] you talk about what you experienced and what did happen […]. [Y]ou don't guide them in the sense that you tell them 'This happened here, this happened there,' but you kind of marry the personal […] to the actual historical. So, so basically […] you weave a tapestry of […] individual human experience […] in the place where it actually happened" (Interview with Gutter 11 September 2020, 1:04:04–1:05:40).

While guides try to incorporate individual experiences of the victims of Nazi terror into their tours, Pinchas combines historical, factual knowledge with his own experiences. For viewers, that creates an emotional closeness to him (Rothstein 2020, p. 217 f.). Therefore, his virtual *guided tour* is more tangible and much more personal for the viewer. Hence, the term survivor-guide is appropriate for describing Pinchas' role in *The Last Goodbye* when one keeps in mind the differences between a survivor-guide and guides at memorial sites. The Neuberger incorporated this approach of guiding into its work in the field of Holocaust education for a specific period, between the October 26 and November 26, 2018: the Centre offered students the opportunity to watch the film in a virtual reality studio (Phillips, personal communication 6 January 2021).

5   Conclusion

Guiding is not a static function, and the work of the guides at the Neuberger demonstrates the variety of approaches to it—this is already reflected on a linguistic level by referring to the volunteers as educators rather than guides. Most notably, two distinctly different methods can be distinguished. "[T]he 'sage on the stage'" method means that the guide imparts knowledge by presenting the information in the form of a monolog. In contrast, the "guide on the side" (King 1993, p. 30) strategy relies on the guide taking a more restrained role—participants of the guided tour are encouraged to acquire historical knowledge by themselves through discussions, research on their own, and through information that is already given in the exhibition. Through this approach, the *Historical Thinking Concepts* are also partly incorporated into the guide tour. Aside from these two methods, self-guiding and accompanying are also applied during a guided tour. As Audrey Diamant explained, all of these approaches are combined when conducting a guided tour at the Neuberger.

Apart from these approaches to guiding, the guides perform yet another vital task—they act as intermediaries between the Centre and the Holocaust survivors who speak at the Neuberger. Intermediaries is probably too neutral of a term since true friendships develop between the guides and the survivors. The Neuberger does not work with a fixed program to train the guides, instead it relies on a personalized approach based on the individual pre-existing knowledge of the guides. A special focus during the meetings of the Neuberger staff with the guides is on the pedagogical techniques of how to impart knowledge about the history of the Holocaust. The personal connection of the guides to the topic and also the close relationships with the survivors who speak at the Neuberger creates an environment in which the informal training-program is eminently suitable. Each of the Neuberger's museum educators conducts tours out of personal interest and concern—they will acquire knowledge autodidactically in addition to what they already know because they want to be able to educate others as much as possible. Therefore, it is sufficient to train them with slightly more focus on how to impart knowledge than purely on historical facts. The fact that the museum educators are volunteers creates a flexible scope that allows for a variation of the guiding techniques in contrast to museums and memorial sites with employed guides. At the Neuberger, varying the guiding techniques is possible, also because most visitors are students. The possibility of speaking to a survivor directly after the tour makes the previously acquired historical knowledge more tangible—the students are not only presented with facts but are able to connect the facts with the personal experience of someone who actually witnessed the Holocaust. Even if the students do not talk to a survivor, the personal connection of the guides creates the same effect, although slightly minimized.

In addition to speaking at the Neuberger, survivors can also assume the role of a *survivor-guide*. While this special method of guiding is applied at actual memorial sites and museums, *The Last Goodbye* made experiencing this method of guiding independent of time and space—although showing the film requires a very specific equipment. The term survivor-guiding not only refers to the fact that the guide is a Holocaust survivor, it also means that historical facts are connected to the personal experiences of the survivor—the guided tour becomes a shared experience of survivor and visitors. However, as Pinchas pointed out, in his opinion "the role of the Holocaust survivor is not to be a guide" (Interview with Gutter 11 September 2020, 1:04:00–1:04:20). Therefore, it is vital to distinguish between the perception of those who listen to him and for whom he is at times like a *survivor-guide* and his own point of view. Either way, survivors accompanying groups at memorial sites are yet another approach to guiding which includes more emotionality due to the personal experiences of the survivor.

Data Base

Interview with Audrey Diamant [Zoom], 07 August 2020.
Interview with Pinchas Gutter [Phone], 11 September 2020.
Interview with Carson Phillips [Zoom], 30 July 2020.
Conversations with Carson Philipps, 20 August 2020, 06 September 2020, 06 January 2021.

References

Sarah and Chaim Neuberger Holocaust Education Centre (ed.). (n.d.). *Teacher's Resource Guide: Brady Family Resource Kit.* Online Version. https://d296fcdc-4762-4d2f-b5a 2b4c0a5e838fb.filesusr.com/ugd/e8cc4e_cd64e52599c94d65a2fdc4871583bfba.pdf. Accessed 01 April 2021.
Ballis, A. (2021). Memories and Media—Pinchas Gutter's Holocaust Testimonies. In A. Ballis et al. (eds.), *Interaktive 3D-Zeugnisse von Holocaust-Überlebenden. Chancen und Grenzen einer innovativen Technologie* (147–166). Braunschweig: Eckert. Dossiers 1. urn:nbn:de:0220–2021–0017.
Bergen, D.L. (2016). *War and Genocide. A Concise History of the Holocaust.* Lanham: Rowman & Littlefield.
de Jong, S. (2018). *The Witness as Object. Video Testimony in Memorial Museums* (Museums and Collections Vol. 10). New York: Berghahn Books.
Gutter, P. (2017). *Memories in Focus. The Azrieli Series of Holocaust Survivor Memoirs.* Toronto: The Azrieli Foundation.
King, A. (1993). From Sage on the Stage to Guide on the Side. *College Teaching*, 41(1), 30–35. www.jstor.org/stable/27558571. Accessed 01 April 2021.
Körte-Braun, B. (2013). Erinnern in der Zukunft: Frag das Hologramm. *Yad Vashem. E-Newsletter.* www.yadvashem.org/de/education/newsletter/10/holograms-and-remembrance.html. Accessed 01 April 2021.
March of the Living Holocaust Survivors Database 1988–2019. (n.d.). International March of the Living. www.motl.org/survivors/. Accessed 19 January 2021.

Ministry of Education, Ontario (ed.). (2018). *The Ontario Curriculum, Grades 9 and 10: Canadian and World Studies, Geography, History, Civics (Politics)*. Ontario: Queens Printer. www.edu.gov.on.ca/eng/curriculum/secondary/canworld910c urr2018.pdf. Accessed 19 January 2021.
Rothstein, A.-B. (2020). *The Last Goodbye*. The First Encounter: Begegnung mit Erinnerungen im virtuellen Raum. In A.-B. Rothstein, & S. Pilzweger-Steiner (eds.), *Entgrenzte Erinnerung. Erinnerungskultur der Postmemory-Generation im medialen Wandel* (193–222). Berlin: De Gruyter.
Rubenstein, E., & March of the Living. (2020). *Witness. Passing the Torch of Holocaust Memory to New Generations*. Toronto: Second Story Press.
The Last Goodbye. (n.d.). Museum of Jewish Heritage—A Living Memorial to the Holocaust. https://mjhnyc.org/exhibitions/the-last-goodbye/. Accessed 26 January 2021.
UJA Federation of Greater Toronto, & Sarah and Chaim Neuberger Holocaust Education Centre (eds.). (2019). *The Holocaust and Now: Holocaust Education Week 2019 Program*. https://cdn.fedweb.org/fed-35/2/HEW2019_program-with-schedule_LR_Rd.pdf. Accessed 01 April 2021.
UJA Federation of Greater Toronto, Sarah and Chaim Neuberger Holocaust Education Centre, & USC Shoah Foundation (eds.). (2018). *Pinchas Gutter Biography*. https://iwitness.usc.edu/sfi/Documents/Canada/TLG_Pinchas_Gutter_Biography.pdf. Accessed 01 April 2021.
USC Shoah Foundation. (2018). *Dimensions in Testimony: One Sheet*. http://sfi.usc.edu/dit. Accessed 01 April 2021.
Watercutter, A. (2017). *The Incredible, Urgent Power of Remembering the Holocaust in VR*. *Wired*. www.wired.com/2017/04/vr-holocaust-history-preservation/. Accessed 01 April 2021.

Links

https://www.holocaustcentre.com/educator-development. Accessed 26 January 2021.
https://historicalthinking.ca/about-historical-thinking-project. Accessed 26 January 2021.
http://historicalthinking.ca/cause-and-consequence. Accessed 26 January 2021.
http://historicalthinking.ca/ethical-dimensions. Accessed 26 January 2021.
http://historicalthinking.ca/historical-perspectives. Accessed 26 January 2021.
https://www.bradyresourcekit.ca/object-2-1. Accessed 26 January 2021.
https://www.bradyresourcekit.ca/object-3-9. Accessed 26 January 2021.
https://www.bradyresourcekit.ca/object-4-8. Accessed 26 January 2021.
http://historicalthinking.ca/historical-thinking-concepts. Accessed 26 January 2021.
www.hanassuitcase.ca/?p=19. Accessed 26 January 2021
https://www.hanassuitcase.ca/. Accessed 02 April 2022.
www.holocaustcentre.com/about-us/who-we-are. Accessed 26 January 2021.
www.holocaustcentre.com/educators-students/field-trip. Accessed 26 January 2021.
www.bradyresourcekit.ca/george-brady-kit. Accessed 26 January 2021.
http://holocaustlifestories.ca/wp-content/uploads/2017/08/Biography-George-Scott.pdf. Accessed 02 April 2022.

The Impact of Digitization on Tour Guiding

A Case Study on Interactive Biographies in Museums

Anja Ballis

Abstract

In this article, I focus on tour guides in light of digitization. Of special concern is to what extent the guides' (called docents in the US) understanding and practices of tour guiding at Holocaust museums vary because of technological innovations integrated in the exhibits. Therefore, I investigated guides' practices with the project Dimensions in Testimony, which the USC Shoah Foundation has been developing since 2011. Three research questions are of interest: To what extent does the institution train the docents for using the interactive biographies? What practices do docents use when presenting the interactive biographies? How do docents evaluate this digital tool for their tour guiding narrative? In order to answer these questions, I use a qualitative approach from the social sciences with interviews and field notes. The results reveal that the guides need technical skill to run the testimony. In addition, they must activate and motivate visitors using the installation. Despite all technological innovations at museums, my study stresses the docents' ability and willingness to connect with visitors on a professional level.

A. Ballis (✉)
Institut Für Germanistik, Ludwig-Maximilians-Universität München, Munich, Germany
e-mail: anja.ballis@germanistik.uni-muenchen.de

1 Digitization and Tour Guiding

In the last twenty years, many Holocaust memorials and museums have been making progress to enrich their programs with digital elements and formats. For example, at the United States Holocaust Memorial Museum in Washington a "Virtual Tour for Students" is accessible, which takes visitors through the exhibit by way of nine stations (www.ushmm.org/teach/teaching-materials/primary-sources-collections/virtual-field-trip/virtual-tour-for-students). The 360° image for each station is combined with worksheets and lesson plans for the students. In a similar way, the St. Louis Kaplan Feldman Holocaust Museum provides a virtual tour through the museum's exhibit and its grounds (https://stlholocaustmuseum. org/about-us/visit/virtual-tour/).

Thus, we see that digitization is not a new phenomenon at Holocaust memorials and museums. Many digital projects are centered around preservation and educational purposes (Ebbrecht-Hartmann 2020, p. 4). Not all of the projects were able to use the potential of the technology, and many educational resources are criticized as being conservative, predictable and uninspiring in content and form (Kansteiner 2017, p. 318). At the present time, we realize that digitization has an enormous impact on the reception process. As Pinchevski states, the first generation of media testimony was strictly dedicated to preservation; in a second phase, we combined preservation with reception. Currently, we are concerned about reception—more precisely, with interaction as a means for memorialization (Pinchevski 2019, p. 89; Popescu and Schult 2020, p. 142).

Visitors can benefit from digital interactive tools, posting and editing in order to express themselves and participate in the exhibit (Kansteiner 2017, p. 316). These new participatory forms of engagement have an impact on established routines in museums.

In this article, I will focus on tour guides and their perspectives on digitization. Generally speaking, guides are "pathfinders" and "mentors" who help to make the spiritual and geographical environment accessible (Cohen 1985, p. 9). In the context of this paper, a tour guide is a person, usually trained by the institution, who guides groups or individual visitors around venues or places of interest, in this case Holocaust museums. Guides are mainly called docents in Holocaust museums in the US, and they have the task of interpreting the cultural heritage in a meaningful and correct manner (Weiler and Black 2015, p. 60). Of special concern is to what extent the docents' understanding and practices of tour guiding at Holocaust museums vary because of technological innovations integrated in the exhibits.

Therefore, I investigated guides' practices with the project Dimensions in Testimony, which the USC Shoah Foundation has been developing since 2011. The central goal of the project is to preserve the possibility for later generations to have a conversation with a Holocaust survivor. To create this form of testimony or interactive biography, each survivor was asked up to 2000 questions over the course of several interviews, relating to their lives before, during and after the Holocaust. So far, over 25 interviewees representing three experience groups and seven languages have participated (https://sfi.usc.edu/dit/faq). The survivors were filmed from multiple angles as they answered these questions in order to create a 2D or 3D projection at a later point. Visitors can pose questions to a digital representation of a survivor, which are then, with the support of speech recognition software, automatically matched to the appropriate pre-recorded answers. In the US, the USC Shoah Foundation project strives to facilitate a virtual dialogue with the survivor and has become particularly influential in museum programs (Traum et al. 2015).

Several museums in the US and Europe host such exhibits of interactive and digital testimonies of Holocaust survivors. The Holocaust museums in Dallas, Houston, Skokie, and Terre Haute integrate the project in their permanent exhibit where the biographies are mainly hosted in a special theatre (https:// sfi.usc.edu/dit). Some of the museums present them as a special program for a limited time. Sometimes, docents lead the visitors through the presentation; sometimes, the visitors have the opportunity to explore the testimonies on their own.

2   Research Question and Methodology

In this paper, I focus on the role of docents in presenting the project Dimensions in Testimony at Holocaust museums. Three research questions are of interest:

1. To what extent does the institution train the docents for using the interactive biographies?
2. What practices do docents use when presenting the interactive biographies?
3. How do docents evaluate this digital tool for their tour guiding narrative?

In order to answer these questions, I use a qualitative approach from the social sciences. Understanding guides as social actors, I am interested in their everyday routines. Focusing on such routines, I aim to reveal their perspective on guiding and consequently follow the methodology of grounded theory with its central idea of anchoring research in reality. One methodological pre-assumption stresses the role of groups investigated. Researchers select various groups to compare diverging or similar evidence. Their goal is to obtain a broad range of acceptable indicators for the categories and properties of the analyses (Glaser and Strauss 2008, p. 49). The freedom to compare various groups follows the criterion of theoretical relevance to systematically generate theory, thus controlling data collection. One of the researcher's challenges is to collect ample data in a meaningful way (Glaser and Strauss 2008, p. 52).

In the summer of 2018, I started my research project by visiting the Holocaust museum in Skokie twice. Taking part in several presentations, I could explore the docents' role in connecting visitors and digital survivors. In Skokie, the visitors explore the testimonies in a Holographic Theater, which is part of different tours at the museum. For example, the Take-A-Stand-Tour deals with human rights education, offering different examples in several galleries. It is directed to 7th graders and above. Further, I interviewed docents ($n = 3$) and museums staff ($n = 2$) to gain insights into the museum's policy and docents' view of their role. Finally, I studied guidelines published for the docents and the structure of exhibit, which deliberately integrates interactive biographies in its program. In doing so, I drew a multi-perspective picture of the docents' practices concerning tour guiding and digital tools. To strengthen the results, I contrasted the sample with empirical research in museums, where visitors can explore the project Dimensions in Testimony on their own, presented in a special exhibit. In 2018 and 2019, I visited the project's presentation at the Museum of Jewish Heritage in New York City and at the Swedish History Museum in Stockholm, observing guides' behaviour and talking to museum staff ($n = 3$). The sample was completed by field protocols of a visit to Los Angeles at the USC Shoah Foundation (summer 2018). Here I had conversations with project managers and educational staff.

The data were transcribed, and I coded them in an open qualitative way according to the research questions (Strauss and Corbin 1996); the interviews were enriched with analyses of my field protocols (Breidenstein et al. 2013).

3   Results

3.1   Guides as Translators—Visiting the Holographic Theater in Skokie

The location of the museum in Skokie is important when analyzing the data from guides' practices. Up to 50 visitors can take a seat in the auditorium facing a stage with a screen. At the beginning of the presentation, the docent welcomes the visitors and turns the lights off. The performance follows a fixed routine: Firstly, a short film underscored by music and statements of the survivors introduces their story. In this way, context for asking questions for the audience is provided (Heindl 2021, p. 126). After the film, the digital testimony is presented on stage. Light flashes all around the stage and a 3D presentation of the eyewitness "appears." In Skokie, the illusion of a person is created by the technique of Pepper's Ghost.[1] Then the second part of the presentation starts: The docent of the museum collects questions from the visitors and poses them to the installation, using a microphone at a standing desk. After the Q&A-session, the event ends with a reflection on the visitors' impressions and an invitation to continue walking through the Take-a-Stand-Center or through the museum. The average duration of each theater experience lasts 45 to 50 min, with the film taking 7 to 10 min and the Q&A-session taking 30 to 40 min (Take a Stand Center. Docent Tour Guide n.d.) (Figs. 1 and 2).

Currently, the museum presents the story of different survivors, among them seven from the Chicago area: Aaron Elster, Fritzie Fritzshall, Sam Harris, Janine Oberrotman, Adina Sella, Israel Starck, and Matus Stolov. Furthermore, the testimony of Pinchas Gutter from Toronto is on display. He was the first survivor contributing to the Dimensions in Testimonies initiative, and his virtual representation has been the prototype of the project (Ballis 2021a). More recently, the testimonies of Anita Lasker-Wallfisch and Eva Schloss are being shown as well (www.ilholocaustmuseum.org/tas/), although they have no relationship to Skokie or the museum.

Since the presentation at the Holographic Theater is integrated into several tours, the museums provide the guides with information and training material on the different media. The museums staff is well aware that the docents work on a voluntary basis. The volunteers are often retired and have different educational backgrounds (conversation with museum management, summer 2018). On the museum's website, the guides find biographical notes of the survivors' lives. Moreover, the guides have access to the transcripts of some of the interviews led by the eyewitnesses. For the Q&A-sessions, a list of top ten questions is

---

[1] The British scientist John Pepper invented this illusion technique in the nineteenth century. In modern versions, a projector installed in the ceiling throws a conventional video onto a reflective floor. This in turn reflects the scene onto a transparent foil or glass plate, which is attached above it at an angle of about 45 degrees. If the angles and distances are correct, an astonishingly three-dimensional and realistic representation can be achieved. This is only true though for the audience located directly in front of the stage. Next to the projection, the appearance would no longer be perceivable to the same extent (Deeg 2017).

Fig. 1 Entrance
Holographic Theater
(photo: private)

Fig. 2 Docent at the desk,
in front of the stage (photo:
private)

uploaded to help them to start a conversation with the audience. Besides providing knowledge of the survivors, these materials explicitly explain the educational goals of the exhibit—and of the museum in general—encouraging the visitors to take a stand for humanity:

> "Ultimately, we hope that students will leave the Take A Stand Center with their eyes opened and their hearts inspired to take a stand for humanity, and live a life of action. If we succeed, we have worked toward the education of the whole student—one who leaves with more questions than answers, a desire to learn more, culturally aware, civically engaged, and empowered to move from KNOWLEDGE → INSPIRATION → ACTION!" (Docent Tour Guide 2018)

Regarding the practices of docents presenting the interactive biographies, the pivot point was the questions. During the sessions I observed, the first challenge for the docent was to get the audience involved in asking questions. Since the docents neither had to explain the technology nor the manner of asking questions, it was their task to find an appropriate way of rephrasing the questions asked by the visitors. Although the film clip ends with the survivor's statement "So, please, ask me some questions," sometimes visitors had to be encouraged. The docents supported the audience, for example, by inviting them "to be brave." One reason for this reluctance is that the visitors had just finished watching the film and were still caught up in its atmosphere. After the light show, the survivor's digital figure appeared on stage, which impressed and astonished the audience. The visitors often expressed their surprise with a "Wow" (Ballis and Gloe 2020, p. 11). Then, it takes a little time before the visitors were able to express themselves. The docents had to make a transition, which enabled the audience to move from a narrative-based experience felt while watching the film to narrative sequences in connecting with the survivor (Nilsson et al. 2016, p. 118). The Q&A-part formed a clear contrast to the previous experience and required more activity on the visitors' side (Ballis 2021b, p. 100). Once the shyness was overcome, docents faced new challenges: They had to adapt the audience's questions to the system in terms of language, content, and technology. The following examples from one session with the Auschwitz survivor Fritzie Fritzshall illustrate the different forms of language modifications:

Example 1
Question by the audience "What was she doing during war time?"
Rephrase of the docent "What work did you do in the factory?"
Example 2

Question by the audience "How did you feel when you were liberated from the camp?"
Rephrase of the docent "Tell us about liberation!"
Example 3
Question by the audience "How did she get in touch with her Dad?"
Rephrase of the docent "How did you come to the US?"

In the first example, the guide specified the general question. Since the docent knew the story of the survivor, he/she could more precisely address the event, thus eliciting an answer for the audience. In the second example, the docent took the opposite approach by broadening the semantic scope of the question. Emotions gave way to a request about liberation. Finally, "relationship" was changed into "moving to the US." The docent anticipated the fact that Fritzie's father immigrated to the US before the war, but at that time, he could not bring his family over. After the war, Fritzie herself came to the US and was reunited with her father. Further, the docents often addressed the survivors directly with "you," whereas the audience chose the third person with "she" or "he" (Gloe 2021, p. 135). To give the docents confidence, the top ten questions lay ready at the lectern.

Fritzie Fritzshall Top 10 Questions

1. How did you cope with being a survivor after the Holocaust?
2. Can you describe what a bystander is?
3. What does it mean to be American?
4. What were your earliest memories of childhood?
5. Can you tell me about going to Synagogue before the war?
6. How did you feel wearing a yellow star?
7. What was it like in the boxcar? (This answer is 11 min long!)
8. Did they tattoo you in Auschwitz?
9. What does resistance mean to you?
10. Can you describe the hunger you felt?

The docents' efforts in translating the questions of the audience and transferring them into the system were closely connected with the self-concept of the museum. The educational team members regard the Holocaust museum as a public space. In their mind, the testimonies of the survivors have to be presented at the Holographic Theater moderated by docents. Thus, they see it as their job to bring about "satisfying experiences," because it is not easy for visitors to ask questions and handle the technological instruments. Moreover, the staff of the museum could observe that many visitors have an accent, or ask several questions in one, or are not familiar with the framing of the testimony (conversation with museum management, summer 2018; Gloe 2021, p. 135). From the museum's perspective, the docents are obliged to ensure a profound and long-lasting experience for the visitors.

The visitor-centered approach has had an impact on the docents' practices. Although they were present, standing at the desk, they tried to stay in the background. In the interviews, they emphazised their technical duties, for example, running the microphone. Further, they avoided eye contact with the visitors; instead "they have eye contact with the hologram." One guide explained that "my style is really to remove myself from the interaction as much as possible" (interview_2_10_2018; Ballis and Gloe 2020, p. 12 f.).

However, the docents did not always fulfil the expectations of the museum management. In my studies, I observed docents evaluating the questions from the audience. They decided whether to ask the questions—or not. In addition, from time to time, the docents answered instead of the survivors and paraphrased the content of a potential answer. Of course, it is appropriate after several failed attempts where the system could not find an answer. Nevertheless, this practice also shows the narrow line for moderators: on the one hand, the audience should have a satisfying experience; on the other hand, the docents should remain discrete and "do their job." One guide describes the crossing of this line:

"I was in there one day, it was the worst experience, and then shortly thereafter an email went out about how to reconfirm. So this docent is answering the questions for the survivor that is not the purpose. […] I was really outraged, that is BASIC you know. This is you don't answer the questions, the hologram answers the questions. You know it was really bad and she must have done it like four times. […] If you are a docent and you know you have been instructed, you know your job is just to rephrase a question so that the technology is able to understand the question" (interview_3_10_2018).

The quote illustrates that the digital experience might conflict with the docents' understanding of tour guiding. The three guides interviewed have created their own approach to the permanent exhibit, anchored in personal interest: The guides decided to volunteer at the museum for different reasons. One was impressed by the survivor Sam Harris (interview_1_10_2018); one travelled to camps in Europe (interview_2_10_2018); another observed tour guides at the museum (interview_3_10_2018). The museum staff supported them to develop their own tour narrative: One docent stressed the role of building bridges between the everyday life of the visitors and the past by asking questions; one docent emphasized photographs in the exhibit which are central for her guiding. She enjoys "unpacking [...] those photographs, being able to ask these kids to think." And she finds that it is "the most wonderful feeling in the world is when they respond" (interview_3_10_2018). One guide compared her approach to reading a novel: She begins in 1933, the first page of the novel, and ends with liberation, the last page of the novel: "I feel you then can build the case very nicely. You can pull your themes, your ideas, and the kids can make the connections back because you have presented a lot of that information" (interview_1_10_2018). Asked how she established such connections, she mentioned the conversations with visitors. She tried to talk about everyday activities, hobbies, and knowledge from school to relate to them (interview_1_10_2018).

The interviewed guides' activities in the Holographic Theater clearly differed from practices of touring they have developed over time. Depending on their own interest in history and media, they looked for opportunities to talk to visitors. Of particular importance for all three guides were the questions they addressed to the visitors. In the Holographic Theater the communication situation is different: The audience—not the guide—asks questions. The digital survivor—not the guide—answers. The guides communicate no longer directly with their group; they find themselves as a medium transferring information between the audience and the digital survivor. These changes in communication behavior are determined by the location of the Holographic Theater. The space with its advanced digital technology has an impact on the interaction and touches the tour-guiding narrative of the docents.

3.2   Guides as Technical Assistants—Using Microphones and Pushing Buttons

The other group I will focus on are guides who came into contact with the Dimensions in Testimony project in the context of a special exhibit. For this purpose, I collected data at the Jewish Heritage Museum in New York City (August 2018) and at the Historical Museum in Stockholm (August 2019). Since I undertook the fieldwork during summer months, I mainly observed individual visitors, often tourists, visiting the museum and stepping in by chance. The average stay of individual visitors ranged from 3 to 15 min. In New York, the testimonies were placed in a separate exhibit room; the survivors Eva Schloss and Pinchas Gutter were on two screens, running at the same time. Text panels were presented at the sides with information about the project and short biographical notes of the two survivors (Shandler 2020, p. 35). At a desk, a microphone enables the visitors to ask questions on their own (Figs. 3 and 4).

In Stockholm, the testimonies were part of the exhibit *Speaking Memories— The Last Witnesses of the Holocaust* in 2019. This exhibit aimed to capture the story of Holocaust survivors living in Sweden today. It is part of the effort to come to terms with Sweden's role during the Nazi regime in Europe. On the one hand, Swedish politics, especially due to Sweden's trade with the German Reich, contributed to prolonging the Second World War, at least from 1943 onward. Historians accuse the Swedish coalition government's actions during the war of indirectly exacerbating the effe+cts of the Holocaust (Ruth 2010). On the other

Fig. 3 Installation at the Jewish Heritage Museum, New York (photo: private)

Fig. 4 Microphone with mouse for input (photo: private)

hand, Sweden has been actively promoting a pan-European culture of remembrance since the turn of the millennium and has made efforts in the education sector to advance Holocaust education (Nietzel 2013, p. 168).

The exhibit consisted of three parts: Loans from Auschwitz, such as suitcases, were presented in a first section. Adjacent was a small, partitioned room that could hold about 20 people where the Dimensions in Testimony was accessible. In Stockholm, the museum staff altered the presentation of Eva Schloss and Pinchas Gutter daily. Although these two survivors have no connection to Sweden, their testimonies were included to enrich the visitors' experience (Scheja 2019, p. 12). Text panels of the project and notes of the biographies in Swedish were placed on the sides (Gloe 2021, p. 140). Further, the visitors could explore the testimony by using a microphone, which was stored in a wooden box. On the box, the visitors found three questions in Swedish and English: "Where were you born? What happened to your family? How did you survive?" The third section portrayed Swedish Holocaust survivors and was the core of the exhibit. Photographs of the survivors were enriched by their personal messages in Swedish and English, reflecting on their survival. In addition, visitors had access to oral history testimonies and could use the database Visual History Archive from the USC Shoah Foundation. Objects of Jewish life were also integrated into the exhibit (Figs. 5 and 6).

Although there are some differences between the installations of the project, the two museums took the idea developed by the USC Shoah Foundation seriously. The foundation's idea is

"to give people the opportunity to have conversational experiences with survivors of the Holocaust and other witnesses to history, far into the future. Each specially recorded testimony enables viewers to ask questions of the survivor and hear responses in real-time, lifelike conversation" (https://sfi.usc.edu/dit/faq).

Fig. 5 Installation at the Swedish History Museum (photo: private)

Fig. 6 Box with
microphone, button and
questions (photo: private)

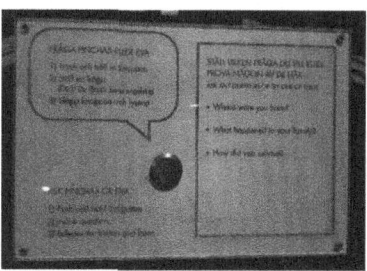

Consequently, the reproduction of the survivors should be life-size and presented at places providing intimacy for the audience. As a project manager of the USC Shoah Foundation argued, the foundation prefers smaller places, bringing the testimonies to the public. Additionally, an integration in the foundation's learning platform IWitness is of special interest because users can meet a Holocaust survivor privately on their computer and explore the testimonies on their own (conversation with management, summer 2018).[2]

Inevitably, the question arises what role guides play in presenting Dimensions in Testimony in special exhibits. For the Stockholm exhibit, the museum management provided their guides with history workshops on the Holocaust and introduced them to the survivors' stories and to the project Dimensions in Testimony. The USC Shoah foundation sent an employee to Sweden who familiarized the guides with some rules for the application. The guides were also instructed how to ask questions to the system. In addition, the employee of the US Shoah Foundation recommended treating the digital testimonies as real people:

"We guide them as if it is a real person. 'So, this is Pinchas, he is here today to meet with us. Hello Pinchas.' 'Hello, who is this?' So, we talk to him like he's a real person. [But] he's a real person that's been filmed. [It] is artificial intelligence. But we decided [to treat it as] a real person, to make it […] a better experience for [visitors]" (interview_management_2019).

---

[2] Since 2020 the interactive biography of Pinchas Gutter has also been accessible on the IWitness website developed by the USC Shoah Foundation. Educators and students are guided through a lesson and provided with biographical information and linguistic advice to help them ask Pinchas Gutter questions. The educational unit on the website—aimed at 14- to 18-year-olds—starts with the history and technical peculiarities of the project. Before the students engage with the story of the survivor, they can read how "[t]echnology allows the possibility for individuals of all ages and backgrounds anywhere in the world to have conversations" (https://iwitness.usc.edu/sfi/Activity/DoActivity.aspx?stp=5c7a2a80-5ed6-4ff6-9ea0-402d4f8f331e).

This idea of a "real person" was formulated into short guidelines for the docents on site. The educational staff prepared a paper written in Swedish with relevant information about the project. They attached a model dialogue in English that seemed appropriate for both Eva and Pinchas being treated as "humans:"

- Hello Pinchas/Eva. How are you?
- Can you please share your resumé with us?
- What is your strongest memory of the Holocaust?
- Can you please sing for us? (To Pinchas)
- Can you tell us about your relation to Anne Frank? (To Eva)
- Did you meet Mr. Mengele?
- How many of your family were murdered?
- What did you remember from the day you were liberated?
- How was your way back to life?
- What can we do to prevent this from happening again?
- What is your most important message for us?
- Thank you for answering our questions today.

In this "dialogue," ten questions are included, so that we find another top ten list. Of importance is the first question "Can you please share your resumé with us?" where survivors deliver a short overview of their lives. In doing so, they provide context for further engagement with their story and help visitors to ask questions.

The practices observed by the tour guides in New York and Stockholm were very similar. Their focus was on explaining the technology and on providing the opportunity to ask questions. At the beginning of each session, they had to motivate the audience for participation, and they instructed them not to ask many questions in one. In Stockholm, language barriers had to be bridged among guides and visitors. The explanations of the technology were in Swedish; the system responded to inputs in English. In both museums, staff members were not regularly present to support the audience. Therefore, it is difficult to decide to what extent their narrative was influenced by the project. However, among individual visitors a reluctance to engage with the digital testimonies could be observed when they were on their own. Only a few visitors took the time to figure out how to elicit answers from the interactive biographies, and only a few used the text banners to find out more information about the eyewitnesses.

In summary, the guides acted as technical assistants, explaining how to use the technical tools, and providing general contextual knowledge to visitors. I did not observe them treating Pinchas Gutter or Eva Schloss as "human beings." Rather, they restricted their comments to the technical specifics of the testimonies and worked with the list of questions.

4   Conclusion—Tour Guiding Between Digitization and Conversation

A comparison of the two groups of guides shows similarities and differences regarding permanent or temporary exhibits of the project Dimensions in Testimony. Both groups need skills in technology: On the one hand, the guides have to be able to run the testimony, as in Skokie; on the other hand, they have to be able to explain the technical requirements to the visitors, as in New York and Stockholm. Further, in both cases, they must activate and motivate visitors using the installation. Not all visitors are comfortable with being a user of technology (Hogervorst 2020, p. 12), even though speech recognition is widely used in everyday life (Duda 2021).

There were also differences between the two groups having to do with the location and the duration of the exhibit: If museums presented the digital testimonies within the setting of an exhibit, an opportunity to interact with these media was created for a limited period. Consequently, the training, the presence of guides on site, and the engagement with the topic were more flexible and lucid.

In contrast, a fixed location, like a theater, created special forms of interaction and communication with which guides had to become familiar. In Skokie, we can see how this form of technology challenged the guides' narratives and self-understanding. The institution curated the theater and integrated the project in several tours. The new technologies heavily influenced dissemination. The docents had to adjust to the requirements of this form of "testimonial landscape" (Wieviorka 2006, p. 116). Up to now, we know little about the guides' perspectives on such forms of digitization and their impact on visitors, especially regarding the audience becoming involved in the creation of memory themselves. The visitors were positioned as listeners who had the opportunity to hear a simulation of someone else narrating or re-actualizing the experience of the past through a recording (Walden 2019, p. 8). Following this line of argumentation, the guides served as another "filter," to unpack Holocaust survivors' experience. Consequently, communication routines have to be re-defined when exploring interactive biographies. New roles of sender and receiver have to be discussed and might become a meaningful educational goal for training guides on site. There should not be interference between the guide and the audience who will ask the question, or between the guide and the survivor who will answer the question.

Nowadays, it is central for tour guiding at Holocaust museums to discuss and reflect on "what you have seen." Conversations on site are still important when touring supported by digital tools. Foundations and institutions develop narratives according to their goals and the structure and capabilities of the media devices. Both visitors and guides react to what they have been presented with. In case of the project Dimensions in Testimony, guides are responsible for guaranteeing a manufactured experience; they are in charge of technical equipment. Of course, they can start conversations with the visitors after the media experience, especially relating to visitors' individual insights—although it might be not that easy because of the immersive effects of the media. Despite all technological innovations at museums, the docents should still connect with visitors on a professional level and prevent themselves from just becoming operators of prefabricated installations, translating content between machine and humans. Docents should become aware of the potential of controversial conversation (Turkle 2015, p. 311), which is crucial to build resilience against hate and which helps to convey an attitude of standing up for humanity. Institutions can encourage their docents to integrate digital tools into their tour narratives combing innovative technology with personal storytelling. This might enrich visitors' reception at exhibits of the Holocaust, and keep the historical event updated, this time in an innovative digital setting.

Data Base

Conversation with Docent, New York, Summer 2018.
Conversation with Museum Management, Skokie, Summer 2018.
Interviews with 3 Docents, Skokie, Fall 2018, interview_1-3_10_2018.
Conversation with USC Shoah Foundation Management, Los Angeles, Summer 2018.
Conversation with Museum Management, Stockholm, Summer 2019.

References

Ballis, A., & Gloe, M. (2020). *"You Have to Become an Upstander!" Holocaust Education und Human Rights Education im digitalen Zeitalter.* https://ulfabraham.de/wp-content/uploads/2020/02/Holocaust-Education-und-Human-Rights-Education-im-digitalen-Zeitalter.pdf. Accessed 31 March 2021.

Ballis, A. (2021a). Interaktive 3D-Zeugnisse von Holocaust-Überlebenden im Deutschunterricht—Theoretische Rahmung, empirische Exploration und disziplinäre Zielhorizonte. In A. Ballis et al. (eds.), *Interaktive 3D-Zeugnisse von Holocaust-Überlebenden. Chancen und Grenzen einer innovativen Technologie* (83–110). Braunschweig: Eckert. Dossiers 1. urn:nbn:de:0220–2021–0017.

Ballis, A. (2021b). Memories and Media—Pinchas Gutter's Holocaust Testimonies. In A. Ballis et al. (eds.), *Interaktive 3D-Zeugnisse von Holocaust-Überlebenden. Chancen und Grenzen einer innovativen Technologie* (147–166). Braunschweig: Eckert. Dossiers 1. urn:nbn:de:0220–2021–0017.

Breidenstein, G., Hirschauer, S., Kalthoff, H. et al. (2013) (eds.), *Ethnographie. Die Praxis der Feldforschung.* Konstanz, München: UVK.

Cohen, E. (1985). The Tourist Guide: The Origins, Structure and Dynamics of a Role. *Annals of Tourism Research* 12, 5–29, https://doi.org/10.1016/0160-7383(85)90037-4.

Deeg, J. (2017). Hologramme. Der Traum von der täuschend echten Abbildung. *Spektrum—Die Woche 18.* www.spektrum.de/news/der-traum-von-der-taeuschend-echten-abbildung/1453825. Accessed 05 April 2021.

Duda, F. (2021). "Was war oder ist Ihre schönste, tollste und angenehmste Kindheitserinnerung?"—Ein sprachwissenschaftlicher Ansatz zur Machine-Learning-Datengenerierung. In A. Ballis et al. (eds.), *Interaktive 3D-Zeugnisse von Holocaust-Überlebenden. Chancen und Grenzen einer innovativen Technologie* (43–62). Braunschweig: Eckert. Dossiers 1. urn:nbn:de:0220–2021–0017.

Ebbrecht-Hartmann, T. (2020). Commemorating from a distance: the digital transformation of Holocaust memory in times of COVID-19. *Media, Culture & Society.* 1–18.

Glaser, B.G., & Strauss, A.L. (2008). *The discovery of grounded theory. Strategies for qualitative research.* New Brunswick, London: Aldine Transaction.

Gloe, M. (2021). Digital Interactive 2D/3D Testimonies in Holocaust Museums in the United States and Europe. In A. Ballis et al. (eds.), *Interaktive 3D-Zeugnisse von Holocaust-Überlebenden. Chancen und Grenzen einer innovativen Technologie* (130–146). Braunschweig: Eckert. Dossiers 1. urn:nbn:de:0220–2021–0017.

Heindl, F. (2021). The Role of Narrative Structures and Contextual Information in Digital Interactive 3D Testimonies. In: A. Ballis et al. (eds.), *Interaktive 3D-Zeugnisse von Holocaust-Überlebenden. Chancen und Grenzen einer innovativen Technologie* (111–129). Braunschweig: Eckert. Dossiers 1. urn:nbn:de:0220–2021–0017.

Hogervorst, S. (2020). The era of the user: Testimonies in the digital age. *Rethinking History,* 24(2), 169–183, https://doi.org/10.1080/13642529.2020.1757333.

Illinois Holocaust Museum & Education Center. (n.d.). *Take a Stand Center. A Docent Tour Guide.* Self-Publishing.

Kansteiner, W. (2017). Transnational Holocaust Memory, Digital Culture and the End of Reception Studies. *The Twentieth Century in European Memory,* 305–343.

Nietzel, B. (2013). Die Internationalen Holocaust-Konferenzen 1997–2009. Von der Londoner Goldkonferenz zur Theresienstädter Erklärung. In: J. Brunner, C. Goschler, & N. Frei (eds.), *Die Globalisierung der Wiedergutmachung. Politik, Moral, Moralpolitik* (149–174). Göttingen: Wallstein.

Nilsson, N.C., Nordahl, R., & Serafin, S. (2016). Immersion Revisited: a Review of existing Definitions of Immersion and their Relation to different Theories of Presence. *Human Technology,* 12(2), 108–134.

Pinchevski, A. (2019). *Transferred Wounds: Media and the Mediation of Trauma*. Oxford: Oxford University Press.

Popescu, D.I., & Schult, T. (2020). Performative Holocaust commemoration in the 21st century. *Holocaust Studies*, 26(2), 135–151, https://doi.org/10.1080/17504902.2019.157 8452

Ruth, A. (2010). *Mythen der Neutralität. Wie der Holocaust in Schweden und der Schweiz ausgeblendet wurde*. www.bpb.de/geschichte/zeitgeschichte/geschichte-und-erinnerung/ 39796/mythen-der-neutralitaet?p=all. Accessed 01 April 2021.

Scheja, L.O. (2019). Opening Words. In *Speaking Memories. Förintelsens Sista Vittnen. The last Witnesses of the Holocaust* (11– 13). Stockholm: J! Judisk kultur I Sverige.

Shandler, J. (2020). The Savior and the Survivor: Virtual Afterlives in New Media. *Jewish Film & New Media: An International Journal*, 8(1), 23–47.

Strauss, A.L., & Corbin, J.M. (1996). *Grounded theory. Grundlagen Qualitativer Sozialforschung*. Weinheim: Beltz Psychologie Verlags Union.

Traum, D. et al. (2015). New Dimensions in Testimony. Digitally Preserving a Holocaust Survivor's Interactive Storytelling. In H. Schoenau-Fog et al. (eds.), *Interactive Storytelling: 8th International Conference on Interactive Digital Storytelling* (269–281), Cham: Springer.

Turkle, S. (2015). *Reclaiming Conversation. The Power of Talk in the Digital Age*. New York: Penguin Press.

Walden, V.G. (2019). What is "virtual Holocaust memory"? *Memory Studies*, 1–13.

Weiler, B., & Black, R. (2015). *Tour Guiding Research: Insights, Issues and Implications*. Bristol: Channel View.

Wieviorka, A. (2006). *The era of the witness*. New York: Cornell University Press.

Links

https://iwitness.usc.edu/sfi/Activity/DoActivity.aspx?stp=5c7a2a80-5ed6-4ff6-9ea0-402d4f 8f331e. Accessed 31 Januar 2021.

https://sfi.usc.edu/dit. Accessed 31 Januar 2021.

https://sfi.usc.edu/dit/faq. Accessed 31 Januar 2021.

https://stlholocaustmuseum.org/about-us/visit/virtual-tour/. Accessed 02 April 2022.

www.ilholocaustmuseum.org/tas/. Accessed 31 Januar 2021.

www.ushmm.org/teach/teaching-materials/primary-sources-collections/virtual-field-trip/vir tual-tour-for-students. Accessed 31 Januar 2021.

Guiding at the Durban Holocaust and Genocide Centre, South Africa

Brenda Gouws and Johan Wassermann

Abstract

The Durban Holocaust and Genocide Centre (DHGC) is one of three South African Holocaust centres under the auspices of the South African Holocaust and Genocide Foundation (SAHGF). The rationale behind teaching the Holocaust in post-conflict environments is that it enables a more open, objective, less emotion-filled examination of a country's difficult past. This is certainly true for post-apartheid, post-colonial South Africa, where violence, intimidation, fear, discrimination, xenophobia, and murder are seared into the country's collective consciousness in both the past and the present. In this paper, we turn our attention to the narrative of the museum educators, who are responsible for guiding all visitors through the permanent exhibition at the DHGC. They are the fulcrum of the educational programme, determining its success or failure. Who are these museum educators, why do they volunteer, how do they acquire their content knowledge and pedagogy, and what do their personal stories mean to their guiding? These questions are answered with the help of qualitative interviews. The paper provides insight into the social background

B. Gouws (✉)
Kaplan Centre for Jewish Studies, University of Cape Town, Cape Town, South Africa
e-mail: bgouws@iafrica.com

J. Wassermann
Groenkloof Campus, University of Pretoria, Department of Humanities Education, Hatfield, South Africa
e-mail: johan.wassermann@up.ac.za

and qualifications of the guides; further, it points to challenges when guiding visitors concerning their languages, their interests in politics, and their schedules on site.

1   Narratives of Guiding

The Durban Holocaust and Genocide Centre (DHGC) is an educational and memorial centre that teaches primarily about the Holocaust but also about other genocides, and claims education as its priority (Pennington 2018, p. 607). To this end, prior to opening its doors, museum educators[1] were sought to present the educational programme to school learners, adult groups, and other visitors to the DHGC. To better understand guiding at the DHGC, this chapter examines three narratives: the narrative of the place; the narrative of Holocaust education; and the narrative of those who do the guiding, both as professional representatives of the institution and as individuals with their own stories to tell. These narratives are graphically depicted in Fig. 1 below.

This chapter is based on a case study that examined Holocaust education at the DHGC and the work of its museum educators (Gouws 2011). The DHGC is one of three South African Holocaust centres, the others being in Johannesburg and Cape Town, under the auspices of the South African Holocaust and Genocide Foundation (SAHGF). The rationale behind teaching the Holocaust in post-conflict environments is that it enables a more open, objective, less emotion-filled examination of a country's difficult past. This is certainly true for post-apartheid, post-colonial South Africa, where violence, intimidation, fear, discrimination, xenophobia, and murder are seared into the country's collective consciousness in both the past and the present.

The DHGC identifies as an educational centre, rather than a museum, meaning its aims are education and memorialisation. Education is at the forefront of its activities and its vision, to create "a more caring and just society in which human rights and diversity are respected and valued," is embedded in its mission statement:

"[To] raise awareness of the evils of genocide with a particular focus on the Holocaust and the 1994 genocide in Rwanda; to serve as a memorial to the six million Jews who were murdered in the Holocaust and all other victims of Nazi Germany and the

---

[1] We use the term museum educator to encompass the dual functions of guides and facilitators. The differences are discussed more fully later in the chapter.

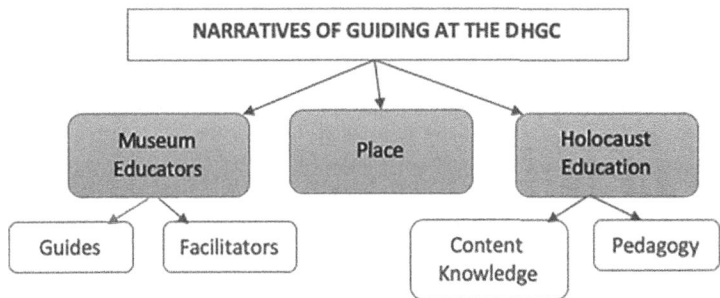

Fig. 1 Narrative of guiding at the DHGC

genocide in Rwanda; and to teach about the consequences of prejudice, racism, anti-semitism, homophobia and xenophobia, and the dangers of indifference, apathy and silence (Durban Holocaust and Genocide Centre 2018, p. 1).

Being part of a global community of Holocaust museums, the DHGC reflects the global trend to incorporate morals and values into its educational programmes. The permanent exhibit, like that of a forthcoming exhibit at the United States Holocaust Memorial Museum (USHMM), operates on the principle that "relevance can be found in the history of the Holocaust itself and that a carefully curated retelling of a particular chapter in Holocaust history can attract, mobilize and empower audiences" (Angell Sievers 2016, p. 295). Previously called the Durban Holocaust Centre (DHC), the word genocide was added to its moniker in 2018 to expand the centre's educational footprint. The Holocaust, the primary reason for the existence of the DHGC, is presented as a case study to examine the past, help prevent future genocides, and to provide learners with tools to deal with social problems by encouraging them to stand up for human rights. The move beyond the narrow focus of pure Holocaust history with the inclusion of other genocides, including the Rwandan Genocide, into the educational programme was deemed necessary to keep South African Holocaust educational centres relevant, by firstly, recognising that genocide is not unique to Jews, secondly, by making them attractive to broader audiences, and ultimately enabling them to attract and retain bigger audiences. Also, particularly in South Africa where knowledge about Jews is limited,[2] museum audiences might tire of hearing simply about a

---

[2] There are approximately 65,000 Jews living in South Africa, within a national population of approximately 58,500,000 people. The majority of approximately 48,000 Jews, live in Johannesburg, with 16,000 living in Cape Town and the rest scattered around the country.

Jewish tragedy, and want to hear about African genocides too, such as the more recent genocides in Rwanda and Darfur in Western Sudan.

The primary rationale for the establishment of the DHGC was, however, to support Holocaust education when it was introduced into the national history curriculum in 2007 for all Grade 9 learners and for Grade 11 learners who chose History as a matriculation elective. At that stage, the Holocaust section in the curriculum, the Revised National Curriculum Statement, had a strong social and human rights focus. However, the educational focus of the current 2011 Curriculum and Assessment Policy Statement (CAPS) has shifted to the historical, with a strong emphasis on race. Notwithstanding these curricular changes, the DHGC has retained its original pedagogy and educational aims.

1.1   The Narrative of the Museum Educators

Our attention now turns to the narrative of the museum educators, as referred to in Fig. 1 above, who are responsible for guiding all visitors through the permanent exhibition at the DHGC. They are the fulcrum of the educational programme, determining its success or failure. So, who are these museum educators, why do they volunteer, how do they acquire their content knowledge and pedagogy, and what do their personal stories mean to their guiding?

Structurally, the guiding body at the DHGC consists of two full-time, paid educators, called facilitators, who are assisted by approximately thirty volunteers, referred to as guides. The volunteer guides are a crucial component of the success of the DHGC as visits to the centre are free.[3] They work in rotation and are called when the centre needs them to accompany schools through the exhibition. The premise of the exhibition tour is for the guides to lead the learners through a virtual world of Holocaust history and help them make personal connections with material that they only read about in textbooks. To achieve this, the guides highlight selected panels, give an account of what happened during the Holocaust, and answer questions, thereby mediating the learners' understanding and interpretation. They are also responsible for modulating learners' reactions, ensuring that they do not become overly upset or emotional. Facilitators, on the

---

According to the Jewish Virtual Library: "In 2018, the Jewish community was estimated at 69,300, 0.5% of the total population" (American-Israeli Cooperative Enterprise 2019).

[3] The necessary funding to run the DHGC and its educational programmes emanates from local Jewish trusts, foundations, and businesses, as there is no governmental financial assistance.

other hand, not only guide the learners through the exhibition, they also conduct the introductory and concluding sessions, and are responsible, together with management, for shaping the educational programme.

Since the DHGC opened its doors in 2008, there have been numerous changes of facilitators, resulting in sporadic losses of institutional memory, but a relatively stable set of guides has been retained. This stability has ensured that the core message of the centre has remained consistent, while new facilitators have injected innovation into the educational programme. Currently, there is a strong contingent of older, White, female museum educators, with a few Black women guides a single male guide. Though small in number, the Black museum educators are highly regarded in the guiding hierarchy as they are needed to communicate the DHGC's message and historical content knowledge to the large number of Black learners who visit the DHGC, in line with national demographics.

The current cohort of volunteers includes retirees, ex-teachers, ex-university professors, a judge, and people with family connections to the Holocaust; their reasons for volunteering to guide varies from person to person. Some want to give back to their communities while others wish to contribute to social cohesion in South Africa. For some, guiding is a means to connect with young people, while for a few, guiding is a way to fill empty mornings, although this certainly does not detract from their enthusiasm or effectiveness as guides. The rationale for the recruitment of museum educators at the DHGC has changed over time. At the outset, for practical and possibly ideological reasons, Jewish community members were invited to volunteer, as many had some prior knowledge of the Holocaust, as we demonstrate in the next paragraph, and sometimes emotional or family connections to it, furnishing them with insider knowledge. This contingent still accounts for a relatively large percentage of guides. Currently the racial demographic of the guides in relation to the general population is negatively skewed, but this is being addressed by the centre with its most recent recruitment drive being focussed on museum educators who speak isiZulu, the mother tongue of most Black learners in the province where the DHGC is located.

In terms of their acquisition of Holocaust knowledge and pedagogy, most of the facilitators have taken up their positions as leaders of the educational programme at the DHGC with little or no in-depth Holocaust knowledge. Their training begins when they receive a pack containing the South African Holocaust and Genocide Foundation (SAHGF) teachers' and learners' manuals, which they are required to study independently. New facilitators are also sent on a ten-day workshop for educators at Yad Vashem in Israel to steep them in Jewish culture and Holocaust knowledge, meaning that their personal practical knowledge of the Holocaust is rooted in Yad Vashem's educational philosophy and pedagogy.

Guides are also given the teaching pack to study and shadow more experienced guides, but do not go to Yad Vashem. This type of knowledge acquisition for both guides and facilitators is fundamentally self-directed and is comprised of both formal and informal elements. The formal component is in the use of official manuals, shadowing, and attending visiting speakers' workshops only for guides, while the unofficial component is found in the trainee museum educators' use of books, videos, documentaries, or networking (Shulman 1987, p. 8). The content knowledge of guides can also vary, based on when they became museum educators. When the centre opened in 2008, formal guide training was conducted by the museum educators of the Cape Town Holocaust and Genocide Centre (CTHGC) who had greater expertise, as the CTHGC had opened ten years earlier. However, as education at the DHGC evolved, the centre developed its own educational ethos and management wanted to achieve greater autonomy. Training for guides therefore changed from formal instruction to the current incidental, constructivist learning model in which learning is self-directed (Pennington 2018, p. 607). In fact, this knowledge acquisition model was described by a new museum educator as "self-training." She described the process as follows,

"For me personally, how I learnt was watching other people guiding otherwise I wouldn't have known how to guide [...]. So, I was given [...] books, the *Educator's Manual* as well as the *Learner's Manual*, and I also had to basically learn from those books. I spent, I think it was a month, ploughing through, trying to get an understanding, more of an understanding, of the Holocaust from those books" (Gouws 2011, p. 93).

Further education for museum educators takes place in the form of intermittent workshops by visiting speakers when they come to deliver talks to the local Jewish community. Yet, ultimately, despite the lack of formal training, guiding at the DHGC can be regarded as successful.

1.2 The Educational Narrative

In line with its aims, the DHGC's core educational strategy is to teach the historical facts of the Holocaust, as well as to provide a strong human rights and moral message. This is achieved in the South African context by highlighting the strong parallels between 1933 Nazi Germany and apartheid South Africa, and drawing out the lessons from the Holocaust about man's inhumanity to man, the need for learners to uphold human rights, and to be agents of change in order to prevent future genocides. As there is no fixed curriculum for the museum educators to

follow, they are free to explore, highlight, and present whatever facet of the Holocaust they see fit, provided they interact with the panels when guiding, realise the
aims of the centre, and complete their portion of the tour in the allotted forty-five
minutes (Oshry 2019).

The structure of a guided tour for learners is determined by management, with
input from the facilitators. The tour begins when the group of learners arrive at the
centre and are ushered into a seminar room for a half-hour introductory session.
There a facilitator welcomes them and explains the day's events, after which
they watch a twenty-minute locally created documentary entitled, *The Holocaust.
Lessons for Humanity* (South African Holocaust and Genocide Foundation 2019).
The film provides a concise historical overview of the Holocaust and discusses
the lessons that can be learned from the Holocaust, addressing issues such as
discrimination, racism, prejudice, and human rights. The groups are then split
into two or four, depending on the size of the initial group, for the next forty-
five-minute portion of the tour. Splitting the learners into groups is necessary
as the visiting groups are usually quite large and the exhibition passages are
narrow. Furthermore, smaller, more intimate groups encourage greater connection
between the guides and the learners. The first group remains in the seminar room
with the facilitator, who uses the film the learners have just watched as a starting
point to delve deeper into the socio-historical issues surrounding the Holocaust.
Discussions take place about what can be learned by studying the Holocaust and
what these lessons mean for society today. Meanwhile, the other group is guided
through the exhibition. As there are only forty-five minutes to cover the expansive
history of the Holocaust, the guides stop and focus the learners' attention at
seven or eight panels. Without a prescribed script, each guide chooses the panels
at their own discretion, emphasising what they feel is most appropriate for the
group as they tell the Holocaust story. This individualisation generally reflects
what is meaningful to the guide personally. For instance, they might highlight
the question of refugees, the notion of choiceless choices, collective punishment,
the fragility of democracy, the fate of the Jews, the prevention of genocide, or
even Christian antisemitism.

Teaching in the exhibition portion of the tour is usually teacher-centred,
depending on the extent of museum educators' Holocaust content knowledge
and the learners' pre-tour knowledge. The museum educator is regarded as all-
knowing, with little exchange of knowledge, ideas, self-exploration, or interaction
with the learners (Blaschitz and Herber 2016, p. 269). In addition, the large
amount of material to be covered in a very limited time means that there is little
room for active participation by the learners. Furthermore, discussions about Jewish history are limited by learners' lack of knowledge of Jews or Jewish history,

with most never having met a Jewish person (Kaplan Centre for Jewish Studies and Research 2016, p. 25). The learners' prior Holocaust knowledge is mostly confined to what they have learned in the classroom. A more participatory teaching approach is adopted during the introductory and concluding sessions, when learners are encouraged to participate more actively. After the first session, the groups are given a fifteen-minute break in the garden where they are provided with a snack and a drink, after which the groups swap for the second forty-five-minute session. Finally, the groups are re-united for a concluding session during which time they are debriefed by the facilitator who provides them with an opportunity to discuss what they have seen and felt, and they are challenged to explore what it means to be a bystander, perpetrator, victim, or upstander. The learners are actively encouraged to stand up in the face of injustice and become agents of social change. They then fill in evaluation forms, which the facilitators later read and assess.

For the museum educators, each school tour is a balance between making meaning of the Holocaust for learners and trying to satisfy the demands of the curriculum while balancing the requirements of the curriculum and the museum's ideology. For the learners, however, meaning is ultimately attained from whatever message they take away from their learning experience, although the duration of their altered consciousness is questionable. There is currently no research that suggests that a couple of hours spent at a Holocaust centre will change attitudes in the long-term. Learning about the Holocaust does not even change learners' levels of antisemitism in the classroom (Gordon et al. 2004), although it has been suggested that simply bringing the Holocaust into learners' consciousness has the power to open minds (Eckmann 2010) and get them thinking about racism and genocide (Short 2015).

As an incentive for teachers to participate in the teacher education workshops at the DHGC, they are able to accrue professional development points (Durban Holocaust and Genocide Centre 2018, p. 47). For adult groups, guiding takes place under the auspices of the DHGC's Department of Adult Education and Social Justice Programmes, the aim of which is to respond to racist rhetoric, particularly on social media, by equipping people with "the information and communication skills to speak back with compassion and dignity" (Durban Holocaust and Genocide Centre 2018, p. 53). To this end, the museum educators guide corporates, small business groups, and other civil groups such as nurses and police, as well as conducting teacher development workshops.

1.3   The Narrative of Place

A narrative of guiding at the DHGC would be incomplete without a conversation about the place at which the guiding occurs, as shown in Fig. 1. According to a publication that celebrated the DHGC's ten-year anniversary, soon after its establishment, the DHGC became "an educational and cultural landmark in the city" and it was only "a question of time before it became a physical one too" (Durban Holocaust and Genocide Centre 2018, p. 25). Today, standing in contrast to South Africa's ugly, fractured apartheid past, race-conflicted present, and the torturous history of the Holocaust, the permanent exhibition is housed in a sleek, modern building that incorporates world-class museum architecture. In addition to the primary museum space, there is a double-void library/lecture theatre, offices, a popular café-cum-gift shop, and a lush memorial garden with a sparkling fountain in the museum precinct. The DHGC, like other museums worldwide, is built on a [relatively] grand scale and is typically expensive. Holleran (2018) describes this as a gesture to public memory, but questions if "big architectural projects" are the most appropriate form of storytelling. She feels that they may not be.

The permanent exhibition at the DHGC, which located in the above space, is comprised primarily of informational panels with photographs and text; hence, the chief manner of teaching and learning at the centre is visual. In addition, there are discreet alcoves where visitors can sit and watch survivor testimony or immerse themselves deeper into the Holocaust experience with various inter-active multimedia screens and a glass display cabinet that houses Holocaust-era artefacts. The air-conditioned pathway through the exhibition is quiet and dimly lit, evoking a reverent atmosphere. As it snakes its way through the history of the Holocaust, soft Wagnerian music morphs into background sounds of Hitler's ranting, and Holocaust victims' testimonies, all of which mingle to create a sombre mood, while strategically placed lighting highlights various-sized, mainly black-and-white photographs with accompanying text on the panels.

A unique facet of the exhibition is a simulation of Anne Frank's attic room, one of only a handful in the world constructed with official permission from The Anne Frank Trust. The Anne Frank room at the DHGC provides a "living history" experience for learners, who can venture behind a simulated, hinged bookcase door into a room that is a reproduction of the room Anne Frank occupied in the attic. The museum educators use this room to try and recreate the "aura" of Anne's experience as an authentic moment (Blaschitz and Herber 2016, p. 269), particularly with younger school groups, although this pursuit of aura is regarded as impossible by some researchers. Digan (2010), for instance, argues that with or without the use of authentic historical objects, purposely creating a direct experience of the past is not only impossible, but also undesirable.

It is into this quiet but evocative space of the exhibition that learners from Durban and its urban and rural surrounds are brought to learn about the Holocaust and other genocides. The learner visitor population reflects all facets of South African society from unsophisticated rural learners attending poor government schools to the digital-savvy, wealthy, continent-hoppers of private city schools. The mediators of this educational space, the museum educators, accompany the learners through the exhibition.

2   Untangling Guiding at the DHGC

The museum educators at the DHGC facilitate a professional learning environment and are deeply committed to Holocaust education. As discussed, the museum educators' content knowledge and pedagogy are crucial elements of their guiding and they shape what they teach.

2.1   Agendas—The Institution, Museum Educators, and Visiting Schools

The SAHGF has described itself as "the only national service provider of Holocaust education" (Du Preez 2008, p. 144) and to fulfil its mission to be the expert and official provider of Holocaust education in KwaZulu-Natal, the centre adopts a mentorship role. Teaching the teachers through Holocaust education workshops means that the Centre can plant educational seeds beyond the physical exhibition, for, as one member of the management commented, "We need to teach [the history teachers] how to teach it right" (Gouws 2011, p. 96).

In addition to this institutional agenda, the museum educators have personal agendas when they guide. As discussed earlier, amongst these are the desire to pass on the lessons of the Holocaust, fulfil their desire to teach, enhance their retirement years, be part of a Jewish educational organisation, and to help others, and these motivations are reflected both in what they teach and how they teach it. A sub-text to these aims is a desire to tell their own stories, although, the depth of these stories is limited by the shortage of time. It is known that adults filter their knowledge through their lived experiences, views, and understanding, thereby constructing their own meaning (Castle 2006, p. 123–132), and this is evident in the guiding at the DHGC. Our research has shown that in addition to their training, previous teaching experience, and personal learning, the museum educators' personal stories play a role in how they guide. Their personal stories are used to fill gaps in learners' ability to grasp difficult concepts or illustrate historical events. The incorporation of these stories makes each museum educator's guiding unique, with their stories becoming the lens through which they teach the Holocaust. For instance, considering the context of the DHGC as a product of post-colonial, post-apartheid South African society, all the museum educators in our study highlighted apartheid and human rights during the course of their guiding, as previously mentioned, but many added dashes of their personal experiences to this narrative thereby contextualising the Holocaust (Gouws 2011). This was done based on the museum educators' acquired knowledge, inherent knowledge, expertise, institutional knowledge, socio-cultural knowledge, and personal knowledge. Thus, ultimately, the museum educators teach what they know, so, in our study, Jewish guides laced their guiding narrative with Jewish thought and experience, while a judge incorporated legal aspects of the Holocaust into his. This personal experiential knowledge was derived from their identities, experiences, and cultural and social backgrounds (Clandinin 1985, p. 361), while their practical knowledge grew out of their personal experiences (Watson 2006, p. 525). Teachers' personal practical knowledge therefore relies on both theoretical and practical aspects, but also embodies a range of emotional and moral knowledge.

Finally, the schools also have their own agendas when they send their learners and teachers on a school tour of the Holocaust exhibition at the DHGC. For some schools, the trip is a reward given to selected learners for academic excellence in history, while the accompanying teachers are able to acquire some expert, in-depth knowledge not only about the topic, but also methodologically. For other schools, taking their learners to the DHGC is a step towards moulding learners' values, increasing their sensitivity to world in which they live, engendering empathy, and possibly reducing incidents of bullying or antisemitism. Schools also wish to provide their learners with greater insight into the suffering of others and thereby sensitise them to the suffering of their fellow citizens under apartheid. However, not all history teachers view the DHGC visit as a learning opportunity for themselves and others simply abdicate their teaching responsibility of the Holocaust, leaving the task to the museum's educators and believing that the single trip will provide the learners with enough Holocaust knowledge. It is also possible, though, that the visiting teachers feel intimidated by the topic and, therefore, feel relief that the topic is being taught by professionals. And finally, for some schools, the DHGC is simply regarded as facet of a multi-museum outing that ends at the nearby beach.

2.2   Focus on Pedagogy and Professionalism

It is the volunteers and facilitators who fulfil the DHGC's agenda of memorial, social cohesion, and education. Engaging facilitators who have no prior knowledge enables the management to mould the facilitators' knowledge and pedagogy meaning that they are unlikely to challenge the agenda or status quo of the centre. Previously, challenges to the system by museum educators have not ended well for the museum educators and it is clear that as long as the guides are prepared to arrive, do their work of guiding learners through the exhibition, and then leave, without challenging the status quo of either the pedagogy or methodology, their contribution is highly valued. Whilst this might lead to the suppression of creative development of the educational programme, it does ensure consistency. Innovation generally takes place at the discretion of management.

One of the challenges to the smooth running of the programme is language. Many Black learners who visit the DHGC do not speak English as their mother tongue, and almost all the guides do not speak isiZulu, the language of the majority of people in KwaZulu-Natal. Even those museum educators with a working knowledge of isiZulu struggle. As one guide who often facilitated for the Black rural school groups commented, "My hugest, hugest, hugest challenge is when I guide Zulu-speaking learners through the Centre" (Gouws 2011, p. 139). She explained that having to translate words like antisemitism that do not exist in isiZulu into "hatred of the Jews" every time it is mentioned, and lack of communication when teaching about the Holocaust was difficult and time-consuming.

As discussed earlier, the pedagogy at the DHGC has changed over time and was adopted for various reasons. With the evolution of guiding at the DHGC, it was felt that the different centres operated in different contexts, with many similar but also dissimilar problems. The DHGC training programme for its museum educators was thus soon changed from formal to informal learning, possibly because there were fewer new guides and those who volunteered were generally professionals or experienced teachers in their own right. They were, therefore, highly professional in both learning about the Holocaust and subsequently in teaching it, and they gave mind, heart, and soul to their volunteering. For each of the guides, teaching about the Holocaust was a mitzvah.[4] It was regarded as a way to add to the national conversation about xenophobia, refugees, apartheid, and

---

[4] Mitzvah means commandment in Hebrew, and, in the broadest sense, it means doing good deeds. Doing a mitzvah signifies living in a way that G-d finds acceptable and as a demonstration of positive and moral living. There are 613 mitzvot in the Jewish Torah, and to follow them is part of the core identity of being Jewish.

human rights, to illustrate that White on White violence was also possible, and to show that genocide is possible in even the most highly industrialised nation in the world. One highly contentious issue for learners is the issue of White on White vs. White on Black violence. Learners are often amazed that one group of White people can be intent on destroying another. White on Black, and Black on Black violence is woven into the fabric of learners' sociological vocabulary, it is part of our apartheid DNA, but White on White violence is often unfathomable for learners.

With the South African history curriculum poised to become more Afrocentric, the guides will be required to adapt once again to a new national history curriculum. South Africa is a changing society with a dwindling Jewish community. Furthermore, certain sectors position themselves from a presentist view against Israel, for example, when Israeli academics were excluded from a conference entitled, "Recognition, Reparation and Reconciliation: The Light and Shadow of Historical Trauma" at Stellenbosch University, the Vice Chancellor regretted that they had not attended, but blankly refused to discuss the matter further, leading to suggestions that maybe his refusal reflected the number of Jewish academics at the University (Feinberg 2018). Despite the changing face of the history curriculum, the message carried by the museum educators at the DHGC today has remained consistently focused on humanitarian issues; a message also supported by many history teachers. Educationally, guiding at the DHGC even with an ever-changing national history curriculum has meant that the museum educators stick to the centre's aims and philosophies. They maintain their own levels of content knowledge and focus on the message they want to disseminate. National narratives about racism and apartheid remain an intrinsic part of everyday South African life and, in spite of the instruction to avoid contentious issues, are an integral part of guiding at the DHGC. In fact, in general, how they deal with them remains the responsibility of individual museum educators, and they choose which issues are addressed, and which are silenced. For instance, an issue that constantly arises is the question of how Jews in the present can do to others, namely the Palestinians, what was done to them. This was particularly evident during the previous Middle East war, the Second Intifada. As this contentious topic could lead to heated, prolonged debate, the museum educators have been instructed to acknowledge the question, but not to engage in discussion about it, rather to avoid the issue by saying that the DHGC acknowledges all suffering and would like to see a speedy resolution to the conflict. One guide said that she gives the following response to this question, "We are all brothers and sisters, and the violence in the Middle East should not be happening" (Oshry 2019). In this respect, there is a silencing of the topic. Possibly due to the time constraints, serious contemporary debates such as these are sidelined, even though the nature of Holocaust education invites deep discussion on myriad topics that aside from the contentious Israeli-Palestinian conflict include questions like what were the Jews' religious and political views during World War II, or what did the Jews do to be hated so much? However, as explained, the DHGC's policy is to avoid discussions about controversial topics.

As mentioned earlier, time remains a global challenge at Holocaust museums and centres (Eckmann 2010). With limited time to cover the vast, complex history of the Holocaust, and usually within a single guided visit, the museum educators are required to establish a focus for their guiding and address the main concepts and concerns of the centre's educational agenda. The museum educators must measure the amount of time spent at their chosen panels and balance it against the vastness of the topic, the academic ability, and the emotion displayed by each group, while simultaneously keeping their guiding relevant in the South African context. While teachers testify to a high level of engagement in Holocaust education in their classes, a finding supported by research conducted in the United States of America and United Kingdom (Clements 2006) reveals that emotion plays a large role. Furthermore, the Holocaust itself is a complex challenging topic, because by engaging with difficult knowledge, the Holocaust opens the learners to socio-emotional and behavioural learning, in a situation where cognitive thought is most usually engaged with (Stevick and UNESCO 2018). Another challenge facing DHGC museum educators is to maintain the learners' interest, particularly if they are part of a school group that includes numerous venues on its agenda for the day, the DHGC being just one. Often in these situations, the learners arrive tired after a very early start to their day trip, or are simply intent on reaching their ultimate destination, the beach; in these instances, it is difficult to retain their interest in or enthusiasm for the Holocaust.

In a changing society, challenges inevitably present themselves but can be regarded as either difficulties or opportunities. Learners, for instance, bring challenging diverse attitudes into the learning environment from various quarters, such as home, school, religious affiliation, peer pressure, and social media, which can be used as a teaching opportunity about diversity and multiculturalism during the tour. The success of the educational programme relies on the professionalism, interest, and personal learning of the guides and facilitators and their ability to deal with whatever challenges or issues a group might present.

2.3   Power and Presence of Place

An often-neglected feature in Holocaust education is the role of the building in which the guiding takes place. As described above, the DHGC has morphed and grown, from humble beginnings in a seminar room to the current large, free-standing permanent structure and it continues to develop. The exhibition itself is constantly under review for expansion and update, so, for example, a selective archive relating to the residents of Durban will soon be integrated into the permanent exhibition. This changing face of the museum reflects a dynamic approach to the exhibition space and provides opportunities for the museum educators to grow and integrate new knowledge into their guiding.

The DHGC is an institution that exists outside of First World Europe, in a Third World country, yet it competes with international Holocaust centres both in terms of scale and grandeur. The educational site where the guiding takes place is unashamedly a symbol of wealth in a city where fifty percent of the youth are unemployed and homeless people abound on street corners. The message the building sends is not necessarily one of suffering, a focus that is almost muted by the grandeur, and the ongoing expansion speaks to impressing a global, sophisticated audience in the shape of visiting professors, interns, directors of other Holocaust museums, and movie makers amongst others. This is not unique to the DHGC as Holocaust structures and memorials like Yad Vashem and the United States Holocaust Memorial Museum (USHMM) also ooze majesty. However, in First World countries, museums are intricately woven into cultural life and occupy soaring museum spaces echo traditional castles and palaces. But in a Third World post-conflict, post-colonial country like South Africa, museums generally do not achieve such prominence, except in exceptional cases, like the apartheid museum at Constitution Hill, where the Constitutional Court presides. Yet, the museum educators and visitors appear to be oblivious to the building, and this signifies various things. Firstly, the grandeur of the building might be uncontested by the museum educators if they are accustomed to the kind of wealth inherent in its walls. But by contrast, the portrayal of opulence and power might assail poverty-stricken learners from rural areas and possibly signify or reinforce the old stigmatism of Jews having a lot of power, influence and money.

There is a stark contrast between the sharp pain, suffering, and ugliness of the Holocaust in the exhibition's dimly lit interior and the bright evergreen exterior garden, and sparkling fountain. But there is a more significant, jarring contrast between the sophisticated interior space of the DHGC, and the learners' and teachers' personal experiences of underprivileged schools and lack of infrastructure that might lead them to equate the building with Jewish money. Indeed, this disjuncture between suffering and opulence could be a distraction from the intended message of the exhibition tour and lead to situations where learners are less affected by the content of the exhibition than what it portrays, as in the case of a recent incident that made local news, where learners were photographed displaying the Heil Hitler salute underneath the DHGC sign outside the centre, in clear contrast to what they had supposedly. In fact, the architecture could even echo the African-American philosophy, "How can you tell me this when you live better than I do?" as demonstrated when the Anne Frank room at the DHC was opened in 2007. At the time, a learner commented that the room was not as cramped as her family's living conditions. These phenomena highlight the intrinsic fracture between the building's structure, its content, and the way in which it might be perceived.

3   Implications for Guiding and Concluding Thoughts

From its original designation as the DHC to the nominally inclusive DHGC, a large percentage of the exhibition panels relate to the Holocaust and not to other genocides, reflecting the main purpose of the centre, that is, Holocaust education and memorial. The shift to the inclusion of other genocides reflects the need of the centre to remain relevant, and though there are discussion panels about the nature of genocide at the start of the exhibition and single panels depicting various other genocides at the end of the exhibition the primary focus of guiding at the DHGC is the Holocaust.

Despite the challenges faced by the DHGC guides, their professionalism, consistency, and enthusiasm remain constant, although the question uncomfortably niggles as to whether a single three-hour visit to an exhibition can bring about the societal changes that museums and centres like the DHGC work towards. The short-term benefits are clear in the feedback received from learners and teachers alike, but the long-term benefits remain untested. So, while there are highly enthusiastic, positive responses from learners and history teachers alike in the feedback forms, the feedback needs to be viewed with a measure of discretion. The topic of the Holocaust raises the emotional temperature of visitors and the heartfelt presentations of the material by the museum educators are sometimes pointedly emotive. More research needs to be conducted to ascertain if glowing reports and rapturous comments received are an emotive response to the museum educators' enthusiasm, the horror in the content, or possibly even a result of learners and teachers not wanting to disappoint the experts. Longer term studies are also needed to clarify how long such emotional responses last and whether they will really transform into agency. Further room for research also surrounds the inconsistency of the presentation by guides. Is it a blessing or a curse? Do the diverse approaches to Holocaust education as a result of the lack of formal training result in exciting variations on the guiding by different guides, or can it result in an inconsistent, possibly incorrect message? Certainly, the quality of guiding differs from person to person. As previously mentioned, though, most of the guides have some previous teaching experience or experience working with young people, which enables them to guide young people effectively. These skills are further enhanced by shadowing more experienced museum educators; but whose style they follow, and how closely they stick to other museum educators' historical content knowledge and teaching methods moulds the quality of their guiding techniques. Furthermore, the lack of oversight means that errors occasionally creep in and, with no formal assessment or regular meetings, there is little opportunity for feedback or for guides to air their views or share their expertise in a collegial environment. Consequently, only the loudest voices are heard. More recently, however, monthly meetings for guides have begun to take place, which ultimately adds authenticity to the tour.

Further education and training for museum educators is another critical aspect of guiding and, for the museum educators, this means attending workshops, reading, and other self-motivated study. Ultimately, the tours depend on the expertise and enthusiasm of its museum educators, and they generally guide well because of who they are. As ex-teachers, history buffs, and people committed to expanding their own and their visitors' knowledge, as well as the fact that they are all well-educated and generally have the time and motivation to develop their skills both intellectually and emotionally, which adds depth and breadth to the educational programme, minimal formal training and education is thus currently sufficient. However, with the changing national history curriculum comes a need for change at the DHGC if the centre wishes to continue to support the curriculum. In future, there will be a need for the museum educators to incorporate a wider range of Afrocentric knowledge, to recruit new, younger guides, who might not have any emotional connection to the Holocaust, and who most likely will not have met a Jew in their lives (Kaplan Centre for Jewish Studies and Research 2016). This means that more formal training will probably become necessary at the DHGC. Using the institutional content knowledge and Holocaust pedagogy of more experienced museum educators, a mentorship programme for new guides could be established. Monitoring and performance are also necessary to expand the professional footprint of the guides.

Ultimately, though, guiding at the DHGC as a place is a dynamic and ever-expanding field. The professionalism and enthusiasm of committed guides remains the single greatest asset for the education programme and feeding their content knowledge is not the only necessity. Guides need to be regularly acknowledged for the wonderful volunteer work they do. Listening to their needs and being able to expand their strengths will provide the DHGC with many years of committed, dedicated professionals, who will to guide visitors through the Holocaust exhibition and educational programme.

References

American-Israeli Cooperative Enterprise. (2019). *South Africa Virtual Jewish History Tour. The Virtual Jewish World.* www.jewishvirtuallibrary.org/south-africa-virtual-jewish-his tory-tour. Accessed 01 April 2021.

Angell Sievers, L. (2016). Genocide and Relevance: Current Trends in United States Holocaust Museums. *Dapim. Studies on the Holocaust,* 30(3), 282–295, https://doi.org/10.1080/23256249.2016.1257218.

Blaschitz, E., & Herber, E. (2016). Mediating The Holocaust Past: Transmedia Concepts at Holocaust Memorials and Museums. Paper presented at the *The Politics of Memory and Oblivion, Modes of Transmission and Interpretation/Politika Memorije in Pozabe, Načini izročila in Interpretacije Slovenia.*

Castle, M.C. (2006). Blending Pedagogy and Content: A New Curriculum for Museum Teachers. *Journal of Museum Education,* 31(2), 123–132.

Clandinin, D.J. (1985). Personal practical knowledge: a study of teachers' classroom images. *Curriculum Inquiry,* 15(4), 361–385, https://doi.org/10.1108/S1479-3687(2013)000001 9007.

Clements, J. (2006). A very neutral voice: teaching about the Holocaust. *Educate,* 5(1), 39–49. www.educatejournal.org/index.php?journal=educate&page=article&op=viewFile& path%5B%5D=60&path%5B%5D=56. Accessed 01 April 2021.

Digan, K. (2010). *The just-add-water Holocaust experience: Creating instant historical experiences in Holocaust museums and memorials with authentic historical objects.* Groniek: Historisch Tijdschrift 43.

Du Preez, M. (ed.) (2008). *A place of memory, a place of learning: The first ten years of the Cape Town Holocaust Centre.* Hands-on Media.

Durban Holocaust and Genocide Centre. (2018). *The Durban Holocaust and Genocide Centre: The First Ten Years and Beyond 2008–2018.* Durban: Fishwicks.

Eckmann, M. (2010). Exploring the Relevance of Holocaust Education for Human Rights Education. *Prospects. Quarterly Review of Comparative Education,* 40, 7–16. https://link.springer.com/article/10.1007/s11125-010-9140-z?msclkid=dc6cf4b3b 2711lecb9e2adadd816cf6e. Accessed 02 April 2022.

Feinberg, T. (2018). *Stellenbosch University discriminates against Israeli academics.* South African Jewish Report. Accessed 01 April 2021.

Gordon, S.B., Simon, C.A., & Weinberg, L. (2004). The Effects of Holocaust Education on Students' Level of Anti-Semitism. *Educational Research Quarterly,* 27(3), 58–71.

Gouws, B. (2011). *Investigating Holocaust education through the work of the museum educators at the Durban Holocaust Centre: a case study.* (Master's in Education by Full Dissertation). Durban: University of KwaZulu-Natal.

Holleran, S. (2018). *Shoahtecture   [Critique].* https://jewishcurrents.org/shoahtecture/. Accessed 31 March 2021.

Kaplan Centre for Jewish Studies and Research. (2016). *Attitudes and Perceptions of Black South Africans towards Jewish People in Cape Town, Durban and Johannesburg.* www.kaplancentre.uct.ac.za/kaplancentre/reports. Accessed 31 March 2021.

Oshry, B. (2019). *Discussion about current guiding methods at the Durban Holocaust Centre.* By Gouws, B. (Interviewer).

Pennington, L.K. (2018). Hello from the other side: museum educators' perspectives on teaching the Holocaust. *Teacher Development, 22*(5), 607–631.

Short, G. (2015). Failing to Learn from the Holocaust. In Z. Gross, & S.E. Doyle (eds.), *As the witnesses fall silent: 21st century Holocaust education in curriculum, policy and practice* (455–468). Cham et al.: Springer.

Shulman, L.S. (1987). Knowledge and teaching: foundations of the new reform. *Harvard Educational Review, 57*(1), 1–22.

South African Holocaust & Genocide Foundation. (2019, 2011). *Resource Materials—DVD. Education.* www.holocaust.org.za/pages/material-set-dvd.htm. Accessed 01 April 2021.

Stevick, D., & UNESCO. (2018). *How does education about the Holocaust advance global citizenship education?* www.nck.pl/upload/2018/07/261969e.pdf. Accessed 01 April 2021.

Watson, C. (2006). Narratives of practice and the construction of identity in teaching. *Teachers and Teaching, 12*(5), 509–526, https://doi.org/10.1080/13540600600832213.

The manufacturer's authorised representative in the EU is Springer
Nature Customer Service Centre GmbH, Europaplatz 3, 69115 Heidelberg,
Germany. If you have any concerns regarding our products, please
contact ProductSafety@springernature.com

Printed and bound by CPI Group (UK) Ltd, Croydon, CR0 4YY
24/04/2026
02096312-0003